BRINGING HOME THE COWS

Family Dairy Farming In Stearns County 1853-1986

by Marilyn Brinkman

with Essays by Annette Atkins, Martha Blauvelt and Fred Peterson

**An Exhibition at the Stearns County Historical Society
Stearns County Heritage Center
St. Cloud, Minnesota
October 3, 1987-October 31, 1989**

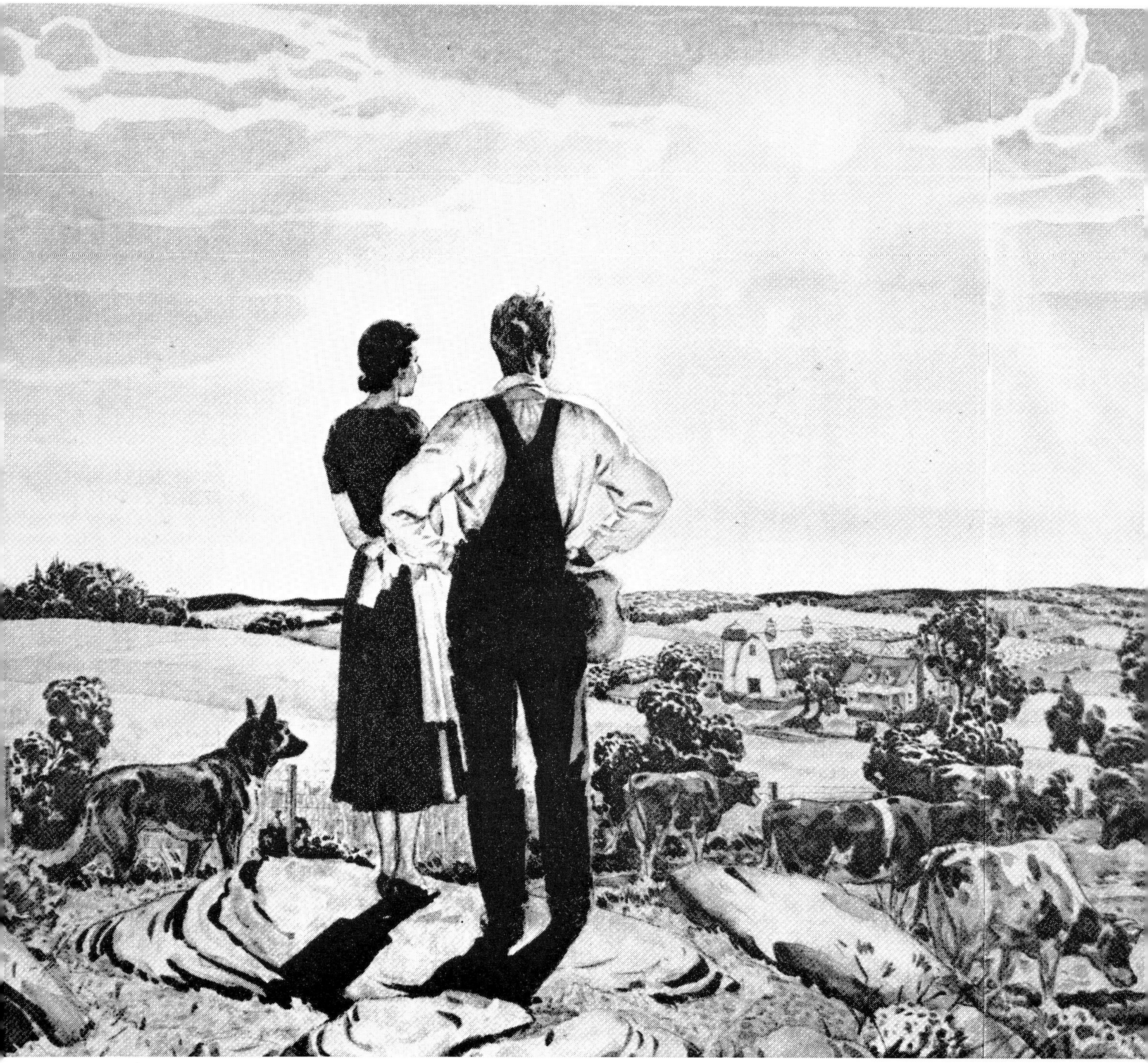

Contributed Articles: **Annette Atkins, Martha Blauvelt, Fred Peterson**
Project Director: **Kevin Britz**
Exhibition Design: **Ann Brown and Kevin Britz**
Photography: **Kevin Britz**
Catalogue design: **Earl Gutnik and Bill Patnaude**
Artifact Catalogue: **Kevin Britz**

Planning Committee: **John Decker, Annette Atkins, Martha Blauvelt, Bert and Gert Schwinghamer, Kevin Britz, Ann Brown**

Executive Director: **David Ebnet**
President: **H. Maureen Beuning**

This exhibition and catalogue is made possible by grants from the *Minnesota Humanities Commission* with the aid of the *National Endowment for the Humanities* and the *Minnesota State Legislature;* and *Bert and Gert Schwinghamer.*

Front cover design: Graphic by **Penny Tennant**, design concept by **Ann Brown** and **Kevin Britz**

©1988 by the Stearns County Historical Society
Library of Congress Catalog Card Number 87-62588
ISBN: 0-929049-00-4

CONTENTS

FOREWORD 5

INTRODUCTION 6

ONE OR TWO FAMILY COWS 7

THE TURN TO DAIRYING 14

UNCLE SAM JOINS THE FAMILY 26

NOTES 38

BRINGING HISTORY HOME/**ANNETTE ATKINS** 41

NEITHER HOUSEWORK NOR WAGE LABOR:
WOMEN'S DAILY WORK/**BY MARTHA BLAUVELT** 44

DAIRY FARM ARCHITECTURE IN STEARNS
COUNTY/**FRED PETERSON** 47

CATALOGUE OF ARTIFACTS 52

FOREWORD

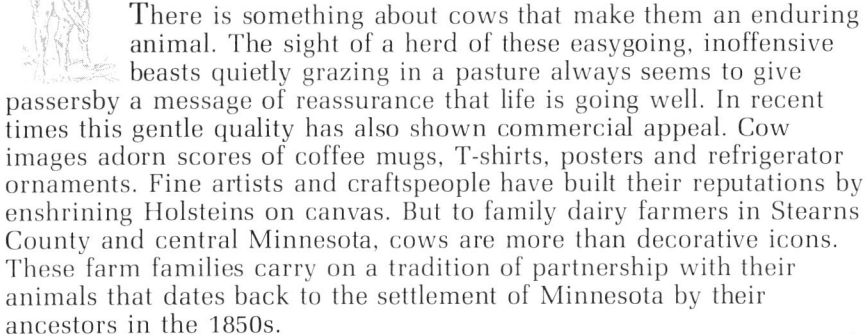

There is something about cows that make them an enduring animal. The sight of a herd of these easygoing, inoffensive beasts quietly grazing in a pasture always seems to give passersby a message of reassurance that life is going well. In recent times this gentle quality has also shown commercial appeal. Cow images adorn scores of coffee mugs, T-shirts, posters and refrigerator ornaments. Fine artists and craftspeople have built their reputations by enshrining Holsteins on canvas. But to family dairy farmers in Stearns County and central Minnesota, cows are more than decorative icons. These farm families carry on a tradition of partnership with their animals that dates back to the settlement of Minnesota by their ancestors in the 1850s.

Perhaps no other animal was depended on more for survival on Minnesota's frontiers by pioneer farmers than dairy cows. The animal provided them with milk, butter, yogurt, cheese, beef, leather, and a beast of burden. The partnership between man and beast never disappeared but continued to grow. Rather than having only one or two cows to meet the needs of their families, today's dairy farmers now manage herds numbering between forty and sixty cows and depend upon the sale of milk as their sole means of livelihood. The process by which the interdependence between cows and farm families has grown is a major chapter in the history of Stearns County and the state of Minnesota.

Stearns County is the largest dairy producer in Minnesota and one of the top ten dairy producing counties in the nation. Despite the ebb and flow of recession, depression, technological revolution, and government incursion, its landscape remains peppered with dairy barns, silos, and pastures of ever-present cows. Some of the farms utilize the ultramodern techniques of computerization, mechanized milk handling, and the advice of paid consultants, others still practice traditional methods passed down from generation to generation. In both, the family remains the heart of the operation.

Because this family industry plays a major role in the lives of the county's people, the Stearns County Historical Society set out to develop an exhibit and catalogue tracing the history of the family dairy farm and the changes brought on by technology, government regulation, and outside economic forces, from the first arrival of cows in 1853 to the present. To compile the history of the industry, we enlisted Marilyn Brinkman, a published historian and local dairy farmer from St. Martin, Minnesota. Her story of the family dairy farm in Stearns County is important both as a primary and as a secondary account.

To illustrate the topic, we collected items within the county, assembling over three hundred artifacts and more than a hundred photographs, making our dairy collection one of the largest in the Upper Midwest. The society relied upon the personal recollections and expertise of many local dairy farmers, creamery operators, farm product dealers, and government ag specialists.

Special thanks to the committee of consultants: Annette Atkins, St. John's University; Martha Blauvelt, College of St. Benedict; John Decker and Ann Brown, Stearns County Historical Society; and Bert and Gert Schwinghamer, Albany, Minnesota.

Those providing ideas and technical assistance included Ann Brown, Lorie Fischer, Sharon Bollig, Robert Lommel, Rene DeRosier, Sr. Thomasette Scheeler, Barb Krebsback, Muriel Generud, Karrie Feist, Dana Dahl.

A special dept of gratitude to Annette Atkins and Martha Blauvelt for keeping the theme on track and the families at heart.

—*Kevin Britz*, Project Director

Marilyn Salzl Brinkman grew up on a dairy farm in Stearns County and witnessed, firsthand, the everyday life of a dairy farmer. Her deep interest in the rural lifestyle has led her to write a number of historical works on the subject. Her recent books include *Light From the Hearth: Central Minnesota Pioneers and Early Architecture* (with Bill Morgan), *The Harvest of Milk: Creameries in Wright County,* and a history of the Stearns County Cooperative Electric Association entitled *Current and Kilowatts.* In addition, she was editor for the *Minnesota Holstein Association* and, as free-lance writer, wrote articles for *The Land* and *Albany Enterprise.*

She and her husband Harold own and operate a dairy farm near St. Martin, Minnesota. She is a graduate of St. Cloud State University and currently works as a researcher for the Minnesota Family Farm Law Project.

BRINGING HOME THE COWS

The first white settlers, predominantly large, conservative, hardworking German Catholic families, brought cows in the 1880s to Stearns County, Minnesota, to supply their families with milk and milk products. The temperate climate, gently rolling hills, numerous lakes, streams, and rivers, and abundant trees and grasses of the area made it ideal for family dairy farming. By the end of World War II dairy farming had become a way of life and the major industry in the county.

The Stearns County dairy industry developed in three phases. The earliest, subsistence farmers, while suffering many hardships, embodied a pure form of family farm ownership and independence. Farm labor was that of the family itself, part of the Jeffersonian ideal in which farmers lived in harmony with nature, with a minimum of outside influence or control. One or two family cows provided milk for drinking, cooking, baking, churning butter, and making cheese. Oxen supplied power to till the land and harvest the crops. This phase lasted until the 1890s, longer in some parts of the country.

The second phase evolved from the first as modern technology (farm machinery) made possible increased production and abundant feed. Creameries provided markets for the sale of milk and milk products. Larger dairy herds were bred and raised to accommodate commercial markets. Farm families moved into the wider world, becoming more dependent on capital and less on family labor. After 1938 rural electricity lightened their work load. World Wars I and II accelerated mass production and capital-intensive farming.

In the third phase, after World War II, government, always somewhat present, became a permanent member of farm families, imposing rules, regulations, and restrictions on dairy operations. Dairy cattle became more productive through advances in breeding, feeding, housing, and disease control, making economies of scale and dependence on capital necessary to economic survival. The independent structure of family farms changed forever. Farmers began to consider themselves part of "agribusiness," the difference between farm families and their urban counterparts grew smaller, and dairy farming as a way of life began to disappear.

Throughout this evolution Stearns County, the top dairy production county in Minnesota and one of the top ten in the United States, has epitomized the dairy industry across the country, yet the family has remained the heart of the dairy industry there.

ONE OR TWO FAMILY COWS
Cows and Subsistence Farmers

Humans have always sought the idyllic way of life. One searcher, Wendell Berry, a contemporary American farmer, writer, and poet, has reiterated Thomas Jefferson's ideal of family farms as "Farms formed by families who live on them, within them, and from them; that farming encompasses a mutuality of influence between farm and household. Farmers own and work the land they live on." Dairy farming, in existence since cattle were first domesticated in Europe and Asia some 6,000 years B.C., supports that ideal in providing the animals necessary for functioning as independent, self-sufficient family units.[1]

Raised and bred in sufficient number, dairy cattle could provide subsistence for a farm family, and the family was the heart of the dairy industry from its beginnings. Early cave and temple drawings and frescoes depict women milking cows. In medieval times, milkmaids—young peasant women—were hired to milk cows and take care of the dairy. The paintings portray children with the women. Men and boys used oxen to perform the heavy labor—planting, cultivating, and harvesting crops.[2]

In colonial America the ideal of family farming and ownership of land began when individual families moved inland away from communities and common property. The cattle they had brought to North America on ships from Europe moved with them on steamboats, railcars, or on foot following wagons as they pursued independent ownership and the Jefferson ideal. The settlers who arrived in Stearns County during the period of heaviest settlement (1853 to the 1890s) were part of this early purist tradition.[3]

Cows and Families on the Frontier

Mainly of German Catholic heritage, the settler families tended to be large, conservative, and hardworking. Retired farmer Herman Kohorst explained, "In Germany if you were born poor, you stayed poor all of your life." In America hard work, land, and cattle could make farmers rich, so people like Phillip Thull, Wakefield Township, "built a long cabin 19 by 26 feet, and started farming with an ox team, a cow and a calf. Sometimes the cow was fastened with an ox and used for draft work."[4]

Cattle were considered great wealth to the settlers. Anton Wartenberg, Paynesville, had two pair of oxen and three cows. Michael Hansen, Cold Spring, arrived with

ONE OR TWO FAMILY COWS

Nineteenth-century farmers with small herds typically milked their cows outside while weather permitted. (Courtesy Minnesota Historical Society).

twenty head of cattle, "possibly the largest herd brought into Stearns County in the pioneer days."[5]

Hardship and deprivation, however, were common among the early farmers: Joseph Primus, Melrose, during "the first winter had nothing in the way of purchased provisions except 100 pounds of flour." His family did not own a cow, so "their principal fare was deer and rabbit meat cooked in various fashions."[6]

Whether there were cattle or not, work on the early farms was difficult and never seemed to end. Men and boys worked in the fields and woods performing the heavy labor, often in adverse weather, with primitive means of transportation and few cleared trails or roads. Women and children did the work of the household, cared for the livestock, and preserved food.

Women cared for the dairy cows, particularly when there were only one or two family cows. They did the milking, churned the butter, fed and bedded the cattle, and cleaned manure out of the barns. The children helped as they grew old enough. Little girls stood with switches during milking time to shoo off flies and mosquitoes.[7]

Boys and girls gathered cows home from the woods and fields at milking time. August Lemke, Albany, told of bells hung around the necks of cattle to make them easier to

ONE OR TWO FAMILY COWS

find. If one family's cattle grazed with others, each family used a different bell. In the evenings when Lemke brought cattle home, he carried a gun to protect himself from bears and wolves. Some families preferred to tether cows on ropes and chains to graze a new circle every day. Milkers simply went out into the pastures with milk stools and buckets to do their work.[8]

This unidentified farm woman used a shoulder yoke when she carried milk from the pasture to the farm home, c. 1880. (Courtesy Minnesota Historical Society).

Butter Making

Milk was used raw as fluid milk, in cooking and baking, made into cheese, or churned into butter—in small amounts for home use or in larger amounts for barter or sale at local stores. Bertha Carpenter, Sauk Centre, made five or six pounds of butter once a week. Adeline Koshiol, Luxemburg, said her mother made and sold butter to the nuns and priests in town. If it had "streaks," it had to be worked longer. In the late 1880s six cents for a pound of butter bought many family necessities.[9]

Although it took time and hard work, the process of

ONE OR TWO FAMILY COWS

making butter was fairly simple: Milk was poured into shallow pans to cool. After the cream separated from the skim, it was ladled off the top and poured into a churn for agitation. When pea-sized lumps formed, they were worked into butter. The by-product, buttermilk, was fed to livestock along with the skim.[10]

Women churned with the help of the children, and grandparents living with the family were expected to help too. Math Ohnsorg, St. Anthony, said he often cranked the wooden churn as a boy, but that he tired out before the butter set and his mother had to finish. Mothers told stories while churning, and it was said that a grandmother could rock the baby in the cradle, hear the younger children's prayers, and churn butter all at the same time.[11]

Farm butter quality and flavor varied from batch to batch and from season to season. Women were proud of reputations acquired for butter commanding good prices in local stores and storekeepers asking for their butter. Many women sold butter formed in special butter molds and prints, often handmade by the men of the family. Poor-quality butter was dumped into large barrels or tubs, from which customers filled their jars, crocks, or firkins.[12]

Because cows did not produce milk all winter, butter was made for winter use in fall and stored in cellars, wells, spring houses, or streams of running water until it froze, or in firkins, tubs, or animal skins, preserved with salt or brine. Often the butter did not keep well or for long. Once the winter supply was used up, the family did without butter until spring.[13]

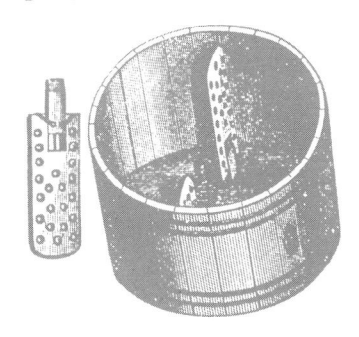

THE HOWE CHURN

Cheese

Early farmers also made two kinds of cheese: fresh and ripened. Ripened cheese was more difficult to make: it had to be processed and aged over time, so small amounts were not practical. Few Stearns County families bothered with it.[14]

Fresh, or cottage cheese, however, was very popular. Rosalia Fuchs, Greenwald, said her family made cottage cheese from milk left on the back of a woodburning stove to sour. The liquid was drained off, salt was added to the remaining curds, and the family ate the cheese on bread or with potatoes. It was often spread on bread and taken to the men working in the fields. Alfred Ebnet, Holdingford, said his mother used another method for making cottage cheese: She put sour milk into a flour sack and hung it from the branch of a tree to drain off the whey. Once drained, the cottage cheese was eaten on potatoes, sometimes with chives added for flavor.[15]

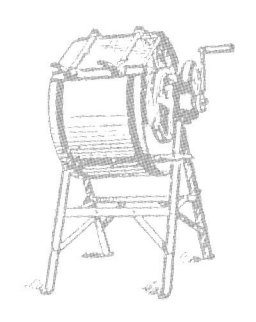

Ice cream, too, was made by nineteenth-century farm families. They packed crushed ice or snow around a pewter or other metal can or jar containing cream and sugar. Then they sprinkled salt over the ice and added water. The salt melted the ice, quickly reducing the temperature of the cream. Vigorous beating and shaking, often happily provided by the children, made the cream swell and freeze. Air incorporated during the shaking process increased the quantity and produced a smoother ice cream.[16]

Ice Cream

Hedwig Winter, Greenwald, said that when she was growing up near New Munich, her parents bought a one-and-a-half-gallon home ice cream freezer, the largest that money could buy. They made ice cream on Sundays, called "feast days" because the older children who worked as housekeepers, priests, cooks, or hired men came home for the day. They flavored their ice cream with vanilla or maple because they "didn't have money to buy fruit." They poured the used salt into the cows' troughs because "We didn't throw anything away on sand land."[17]

Ice cream was only a seasonal treat for most early farm families. In winter, ice was readily available, but abundant feed was not. Cows did not produce enough milk to justify making ice cream.

Work Roles of the Family

Work roles, though somewhat defined by gender, were flexible with the changing seasons and conditions. Responsible for child care, household chores, canning, preserving food, and caring for the dairy animals, many women also did fieldwork, often with small children in tow.

George Kulzer, Albany, wrote of his wife Gretl: "When evening came we walked together, side by side, and would feed the cattle, then sit and rest a little, but mother [Gretl] cooked the meal, then helped with the milking, did the mending, baking, and housework, not quitting until after midnight."[18]

Clara Symanietz, St. Stephen, said she helped with all the work: "Milking was my main job, but when potato-digging season came, I put my baby in a soda cracker box in the field beside me while I picked up potatoes. During the time Vince [her husband] hauled one load to town, I

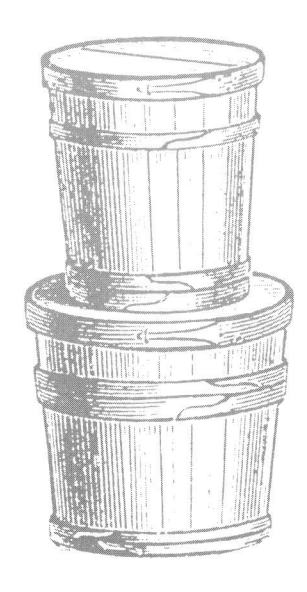

ONE OR TWO FAMILY COWS

picked up another. As the children got older, they were put on blankets, and later they helped with the work."[19]

It was not uncommon for very young children to work side by side with adults in the fields, woods, or barns. Generally, children were given easy tasks until they were strong or old enough (whichever came first), for the harder ones. They often cared for young animals, especially calves. When they helped with milking, they were first assigned easier cows or those almost dry.[20]

By the age of twelve, boys and girls shared milking chores equally with their parents, especially if their herd had grown beyond one or two cows. Daughters helped their mothers care for younger children, while boys relieved their mothers of some of the outside work like carrying water to the cattle.[21]

Heavy outdoor work stopped when darkness came and the men and older boys returned to the house. Many spent long hours after supper, making wooden tools and furniture—even shoes. The Herman Manz family, Paynesville, brought patterns for making wooden shoes from Virchow, Germany. Alexander Korte, a priest born in Farming, wrote in a family history of his father making wooden shoes to wear in the barn. Others made wooden tables, chairs, bowls, hoes, rakes, children's toys, and other useful items.[22]

Despite owning land and cows, many early farm families had a hard time getting ahead or even making ends meet. Often men had to resort to jobs off the farms to supplement their meager earnings. In some families men were gone for weeks or months at a time, to western Minnesota and the Dakotas to work in the wheat harvest fields. Some men spent long winter days in the woods cutting cordwood for home use or for sale. If they sold wood, they drove to the nearest town and often did not return until the next day. The women maintained the farms in their absence.[23]

Not all farm families consisted of father, mother, and children. Single men and women farmed, and some women were forced to farm alone after their husbands died. "Michael Friedman, Cold Spring, died at age 34, and his widow and four children worked at whatever she could find to do, and thus kept the family together." George Kulzer wrote of a family in Avon in which the young mother died: "Five children were left, four boys and a girl, fortunately old enough to do most of the housework." In such instances, children were expected to assume adult responsibilities and contribute to the welfare of the family.[24]

Schools and Churches

It was important to early immigrants that their children learn to read and write in English, so they sent their children to school, usually one-room country schoolhouses begun by area farm families themselves. Many children, most frequently boys, did not attend school when there was work to be done, particularly at planting and harvesting time. Dairy chores, however, could be taken care of before and after school.[25]

Immigrants with similar nationalities and religious beliefs tended to settle together. Religious services were conducted for farm people first in larger homes, then in established mission churches, by traveling priests and ministers. Andrew Gogala, St. Anthony, said they sent messengers ahead to let the people know when the priest would arrive. Women did not attend such services; they stayed on the farms to care for younger children and watch the livestock.[26]

The social life of early farm families revolved around house and barn raisings, church festivities and feast days, and evening visits with neighbors. Families tended to socialize with other families of similar religious beliefs. Children were generally included.[27]

The ideal of living on, with, and from the land changed little for most farm families until the 1890s. Exceptions included single farmers and farm families with money to hire help and those near towns and cities, who had easier access to supplies and other advantages. Some progressive farmers attempted new farming methods, such as winter dairying and commercial marketing of dairy products, advocated by agricultural experts. In Stearns County, old ideas and ideals were not easily relinquished and farmers were slow to give heed to scientific methods of farming. By the 1890s, new events made many farmers of the area willing to consider change.

THE TURN TO DAIRYING
The Rise of the Dairy Farm

The shift away from subsistence farming began when farm families started using steam-powered tractors and other farm machinery. By the 1890s the availability of planting, cultivating, and harvesting machinery challenged Stearns County farmers to adjust their use of land and labor to increase production and ease family drudgery.[1]

Farmers were encouraged to turn pasture and fodder acreage into cropland, so that more feed could be raised for cattle. With more and better feed, dairy cows did not dry up and could be kept producing through the winter, increasing profits. Farmers began to increase their herds. Larger herds, however, produced more milk than most farm families could use themselves or sell locally. Improved methods of storage became a never-ending concern as farmers looked for commercial markets for their products.

In the 1890s, farmers began winter dairying because new technology enabled them to place more land under cultivation and raise enough feed to sustain milking cattle over winter. Cattle like these on the Hansen family farm, Rockville, grew a thicker coat of hair to protect them during the cold Minnesota winters.

THE TURN TO DAIRYING

Creameries

Creameries answered that need. Creameries were private or cooperatively owned businesses that purchased cream from farmers, manufactured it into butter, then sold it to outside markets. As more Stearns County dairy farmers "turned to dairying," more creameries entered the business.

Certainly this change did not occur quickly, nor at the same time or at the same rate throughout the county. In some areas farmers did not wish to sell their milk. The number of cows producing milk on individual farms grew slowly, and many did not have enough cows producing enough milk to sell commercially. Some farmers kept their milk on the farm too long, delivering sour milk and receiving poor prices. Price disputes between farmer and creamery were common. Many farm women were reluctant to let go of butter making as a home industry, and not a few churned butter for home use long after creameries were established.[2]

By 1887 three privately owned creameries operated in Stearns County—two in Sauk Centre and one in St. Cloud. In 1894 those three no longer existed, but there were four others—one each in Richmond, Avon, Melrose, and Sauk Centre. Two years later there were fourteen. By 1913 seventeen cooperatives, fourteen independents, and two centralizers (large corporate butter factories)—thirty-three in all—were in operation.[3]

The interior of the Meire Grove Creamery around 1920.

The Fairhaven Creamery around 1940. The architecture of creameries is often compared to that of churches because of their importance and standing in the community. Their cupolas reached skyward, like church steeples.

Farming, Minnesota, creamery day in about 1910. Butter maker John Michaels (standing in buckboard at right), who lived on the top floor of the creamery with his family, welcomed patrons and their families to the creamery.

15

THE TURN TO DAIRYING

Expansion of Dairy Operations

By the 1890s the John Schwinghammer family, Albany, had increased its dairy herd to eight cows, three two-year-olds, and seven heifers. The family "gradually turned to dairying" as the main farming activity. By 1920 the average herd in Stearns County had eighteen to twenty-five cows.[4]

Most of the cattle were still crossbreeds—general-purpose cattle that the early settlers had brought to provide power, meat, milk, and hides. They had been bred indiscriminately with scrub (poor-quality) bulls, resulting in inferior dairy stock. About this time, though, dairy experts began advocating purebred stock and selective breeding for higher milk production and for a more hardy, healthy animal that could survive Minnesota winters.[5]

But Stearns County farmers were slow to change. In 1920, of 42,543 dairy cows in the county, only 2,400 were purebred. Of the distinctive breeds Holsteins were most numerous, followed by Milking Shorthorns, Guernseys, Jerseys, Brown Swiss, Angus, Herefords, and Red Polls.[6]

With more cattle and an emphasis on milk production, family dairy chores changed. Men and boys began to do many of the winter chores, relieving women of some of the

On cold winter mornings, patrons gathered in the Farming Creamery to socialize with butter maker Ted Niehaus (left), c. 1930.

Isidor Schwinghammer stopped in east St. Cloud to have this photograph taken after showing his cattle at the 1920 county fair. His son Bert sits on the hood of the truck. After 1900 many dairy farmers like the Schwinghammers began using purebred stock to improve the production of their herds.

THE TURN TO DAIRYING

winter drudgery. Children still joined the work force as soon as they were able to do simple chores. Everyone helped with the milking. Art Borgmann, Sauk Centre, said the children in his family began helping in the barn by feeding calves; they progressed to feeding cows, then to milking. Bill Vouk, St. Stephen, said when farm familes had fewer than six or seven cows the women did the milking, but with twenty or so everybody pitched in. William Cooper, Watkins, said his youngest sister was the best milker in the family.[7]

New Technology—New Chores

As families "turned to dairying" after 1900, they began adopting new farm technology. The most important new development was the use of the cream separator, a machine invented in 1878. Cream separators gave farmers a reliable means of separating cream and skim milk at home and eliminated the backbreaking task of hauling whole milk to creameries. The only drawback of cream separators was their need to be washed after each use—a painstaking chore which was added to the work load of farm women and children.

Farm families also started to use milking machines. Available after about 1905, the machines were reputed "to take the place of a hired hand" and to "save from sixteen to fifty-eight man-hours a year per cow." Jerome Pfau, Freeport, said his family bought three Sharpless brand milking units in 1920. They were powered by a battery-operated home power plant. Gerhard Gamradt, Sauk Centre, talked about his family turning to milking machines in 1926, then milking again by hand during the Great Depression. Their unit was powered with an electric plant powered by the wind. When the wind didn't blow, they used batteries.[8]

Milking machines eliminated hand milking but increased the farm women's work load in another way. Every day the machines and separators had to be taken apart for washing. Clara Jenc, Sauk Centre, said washing the dairy equipment was always women's work. Sometimes she hated the work, but as she got older and did fewer farm chores, it became a way of remaining useful.[9]

Ceil Salzl, St. Martin, said the men of her family left immediately after milking to work in the fields or woods, so the women had to wash the dairy equipment: "They were like the dishes in the house, only this was done in the milk room." Children, especially girls, often assumed this chore when they were old enough. They also fed skim

Woman washing a cream separator. Like most farm women, this unidentified woman was responsible for washing the dairy equipment, c. 1935. (Courtesy Wright County Historical Society).

SHIPPERS OF MILK, ATTENTION!
WARREN MILK BOTTLES,

THE TURN TO DAIRYING

One day's sales of DeLaval cream separators from Jacob Feiden's store, Richmond, destined for St. Martin customers, 1917. Cream separators freed farmers from the painstaking task of hauling whole milk to creameries.

milk mixed with grain to the hogs, and they fed hay to the cows and gathered them home for milking. Even very small children liked to dip milk out of cans to take to the house for cooking, baking, and drinking.[10]

World War I: Food Will Win the War

During World War I rising demand and higher prices paid for dairy products by the United States government vigorously stimulated interest in commercial marketing and winter dairying. The government asked farmers to produce the "Food that would win the war," and farmers realized the potential national and international market.[11]

A New Face for the Dairy Farm

Farm architecture reflected the new interest in production. New, larger farm buildings appeared in Stearns County, changing both dairy farm life and the rural landscape. Silos provided prolonged storage without spoilage for corn, grass, legumes, and other plant stalks, and they were easier to fill than barns. Dairy herd size could be adjusted to the production of hay and pasture and to the storage of winter feed. Farmers could carry more cattle over winter. By 1922 there were 1,200 silos in Stearns County. Substantial new dairy barns, often designed by Farm Bureau or county extension agents (people who worked in counties as advisors, representing

THE TURN TO DAIRYING

university "ag" schools), offered huge storage, up-to-date milking facilities, and individual stalls and stanchions for cows. Barn dances were often held in new barns before they were used. All the friends, neighbors, and relatives who had helped with the construction were invited. Music was provided by local people who played musical instruments. Children were included in the festivities.[12]

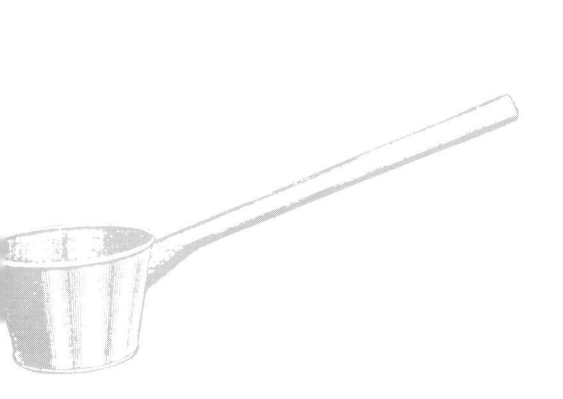

The interior of the H. H. Meyer barn included state-of-the-art cattle stanchions and maximum use of space.

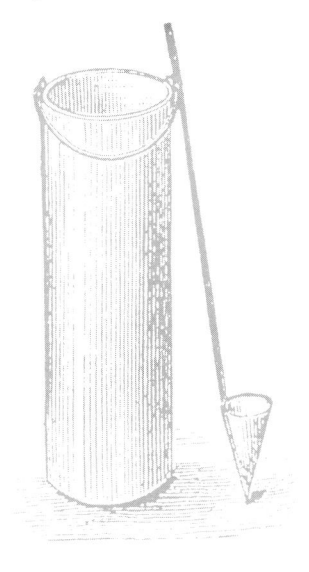

This modern dairy barn on the H. H. Meyer farm, Grove Township, built in 1923, was one of many designed by Stearns County extension agent P. W. Huntemer in the 1920s.

19

THE TURN TO DAIRYING

As farm structures, machinery, and operations became larger and more commercial, work roles became more clearly defined by gender and age. Men and older boys operated the large machines and conducted business in town. Women and older girls and children worked long days as before, but usually at lighter tasks.

Rural Electricity

Rural electricity did more than anything else to alleviate the work load for farm families. People on the East Coast had had electric power since the late 1800s. It spread quickly across the United States into larger cities and gradually into small cities and rural towns, but it was not cost effective for private utility companies to extend lines into rural areas. The Rural Electrification Administration Program (REA), legislated by Congress in 1936, was designed specifically to make electricity available to all rural areas. Rural electric cooperatives were formed, and electricity became available to farmers in Stearns County after August 1938, when the Stearns County Cooperative Electric Association organized in Melrose.[13]

Rural electricity relieved farm families of the drudgery of hand milking, milking with battery- or wind-powered milking machines, and carrying water to livestock. Electrically powered sanitation and sterilization equipment improved safety and health. Farm families could work, read, play, and children could study by electric rather than lantern light.[14]

Time spent on milking chores was cut 45 percent, requiring fewer laborers. Electric coolers and dry refrigeration reduced the rapid multiplication of bacteria that soured milk. Heaters and water pumps provided running water for washing equipment, and water, fresh and cool, was always at hand for drinking. There were more than 200 uses for electricity on the farm by 1947. Winnifred Claude, Sauk Centre, said, "Electricity made us feel we were as good as town people. We could have running water and a bathroom too."[15]

Electricity and refrigeration also added to the enjoyment of family social life. "On the first evening we had power," said Robert Imdieke, Elrosa, "we had homemade sherbet that my mother made. We thought more of the refrigerator than the stove because we just never had cool or frozen food on the farm before."[16]

Not every farmer grabbed at electricity as it became available. Many who did not understand a power that they could not see or hold in their hands preferred to wait until

Electric powered walk-in refrigerator, 1938. Electricity gave dairy farmers greater control of routine milk handling operations.

THE TURN TO DAIRYING

it was "safe." Others, suffering from the Great Depression of the 1930s, simply could not afford it.

The Great Depression

The depression had taken its toll on Stearns County dairy farmers. A drought during the 1930s caused a severe shortage of feed. Many farmers had to sell part of the herds they had painstakingly built. Some farms were lost and many farmers had to turn to the government for relief. Alfred Ebnet said the government permitted bankrupt farmers to keep six cows, twenty sheep, four sows, four horses, and feed. They had to pay interest and fire insurance premiums on farm buildings for three years. Ebnet sold everything upon bankruptcy that he was not permitted to keep. He carried the money from his sale in his shirt pocket for three years because he did not trust banks.[17]

Farm families suffering through the depression became more conservative, their hopes for a good life in America shattered. Those who did not become destitute became more frugal, less likely to gamble on nature and move ahead indiscriminately. Many farmers say that those who have not gone through such a hard time do not understand what real hardship is. Henry and Ceil Salzl, St. Martin, paid $165 for each cow they bought in 1926 when they were married. During the depression they sold some of these same cows to the government for $16 a piece. They were not the only ones: the government bought 13,000 head of Stearns County cattle in 1936, and at least that many were sold privately by farmers. The number of farms decreased from 4,896 in 1935 to 4,685 in 1940 and never again reached the previous number.[18]

Ironically, as the number of cattle decreased, improved nutrition, breeding, and housing resulted in the remaining cows producing more than ever. By 1941 Stearns County was the top dairy-producing county in the state, and the number of milk cows per farm increased steadily until the 1980s.[19]

World War II

On the heels of the depression came World War II. Domestic and foreign markets opened as they had during World War I. American farmers responded by raising 50 percent more food annually on less land with fewer workers than they had during World War I.[20]

Even before the United States entered the war in

The Legatts, a Stearns County dairy farm family on the farm near St. Stephen in 1938.

THE TURN TO DAIRYING

December 1941, the government was promoting increased milk production on America's dairy farms. The War Food Administration established rates of payment to dairy producers and encouraged new farming techniques. The United States Department of Agriculture (USDA) called milking machines "war tools" and claimed they saved 210 million hours of labor per year, important when young men were leaving farms for the armed services or for jobs in factories. Farm families were made to feel like the most important contributors in the war effort.[21]

In 1941, when "Food for Defense" programs were initiated, there were 61,500 milk cows in Stearns County. The annual extension service report for 1941 stated: "The county has the cattle, the feed, the markets, and with good prices, everything is all set to go." Advertisements in the *St. Cloud Daily Times* called the dairy cow "Stearns County's Best Business Woman." Full-page ads were addressed to increased production, and the National Select Service told farm boys, "It is more important for you to stay on the farm and do a good job than to volunteer or be drafted into the armed services."[22]

As young men left farms for the military or to work in towns and cities, younger children had to work at more difficult tasks on the farm. Women, too, were called on to contribute more. Meanwhile, farmers were fighting a war of their own. Although the USDA promoted the purchase of milking machines, the price of machinery and farm supplies, when available at all, went up. Tanks and airplanes were being built instead of milking machines and tractors. Farmers pieced together old machinery when they could not buy new.[23]

New Milk Products

World War II prompted the manufacturing of dry powered milk for human consumption—ideal for troops overseas because it had one-tenth of the bulk of fluid milk, was cheaper, kept for months without refrigeration, and was germ-free because of the intense heat used in processing. In 1941 there were dry milk plants in Melrose, Albany, and Opole, and many creameries were shipping skim and buttermilk to other creameries that could produce powdered milk. By the end of World War II, ten of the top hundred milk-producing counties in the United States were in Minnesota. Stearns County ranked first in Minnesota, twenty-first in the nation.[24]

The promotion of total war production had far-reaching effects on family dairy farms, affecting even their

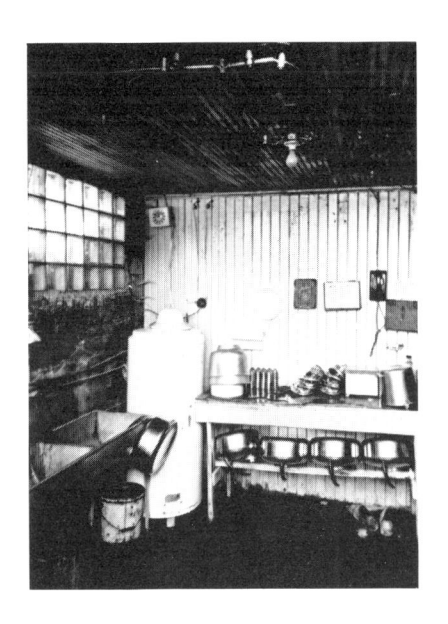

THE TURN TO DAIRYING

definition. After World War II the political definition of family farms was modified to include those farms in which "the family does most of the work, with some hired labor"; also "the basis of the farmer's welfare and independence is no longer landownership but income."[25]

The official definition read: "The essential characteristics of a family farm are not to be found in the kind of tenure, or in the size of sales, acreage or capital investment, but in the degree to which productive effort and its reward are vested in the family. The family farm is a primary agricultural business in which the operator is a risk-taking manager, who with his family does most of the farm work and performs most of the managerial activities."[26]

This grade A milk room on the Hansen brothers farm, Rockville, was typical in Stearns County in the 1950s.

Bulk Tanks and Modern Technology

Through the 1950s and into the 1960s and 1970s, family dairy farms continued to evolve technologically. Bulk handing of milk, farm bulk tanks, elevated milking parlors, loose housing of cattle, pole barns, mechanical barn cleaners and manure pits, sealed silos for storage of high-moisture feeds, bunk feeders, silo unloaders, computerized

This unidentified man is using a stainless steel bulk tank for storage of his milk. By the 1960s tanks like this one replaced cans on most Stearns County dairy farms. (Courtesy Minnesota Historical Society).

THE TURN TO DAIRYING

Charles Hansen of Rockville, milking cows for the last time with Surge machines on October 9, 1961. The Hansen brothers, Pierre and Charles, installed a modern, automated system shortly afterward.

The certificate of registration for Schwinghamer farm Joyce, an award winning Guernsey owned in the 1960s by Bert and Gert Schwinghamer of Albany, Minnesota. The cow was the first Minnesota Guernsey to produce over 1000 pounds of milk per year.

feeding, housing, and breeding, and other advances mechanized and increased production in dairy operations.

Again, the changes did not occur all at once or even in the same area at the same time. In 1955 Minnesota had 2,000 bulk tanks in use; one year later there were 7,200. But a farmer with a dairy operation milking thirty or forty cows mechanically might live just across the road from a farmer milking a few cows by hand. Changes to more modern facilities often came about when a new generation took over the family farm. Still, in 1952 the Stearns County extension agent said, "The production pattern of almost every farm in the county is geared to livestock with dairying in the number one spot."[27]

Specialization in dairy farming in Stearns County resulted in 590 fewer farms in 1956 than in 1936, but these farms had more cows and more production per cow. In 1936 there were 48,183 cows in the county; in 1956 there were 61,000. Family dairy farms had become agribusinesses more dependent on capital than on labor. The family still performed most of the labor, but capital could buy labor and machines to assist when production capacity exceeded home labor capacity. More farm children attended high school, more graduated, and many attended college.[28]

24

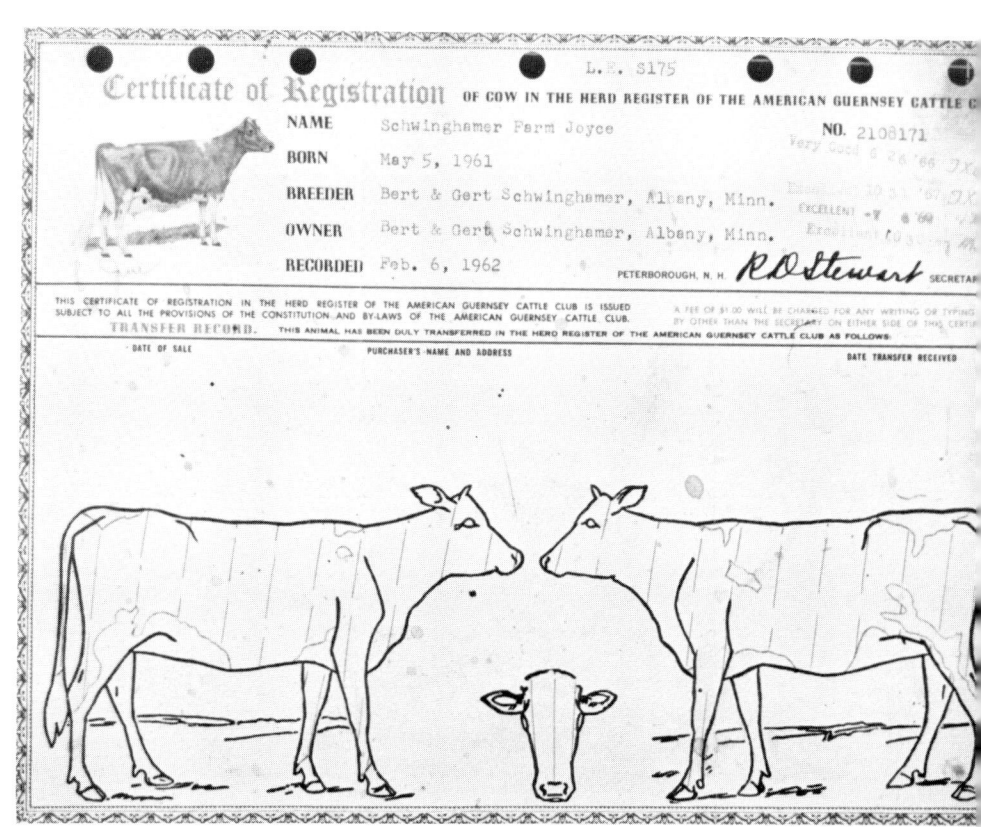

THE TURN TO DAIRYING

Mechanization and government subsidies made it possible for many families to raise their standards of living. Increased dairy production presented opportunities to farm children never before possible. No longer bound to work on the farm, children could leave. Cities offering high wages, college education, and other advantages gave them a place to go.

The social lives of farm families changed. Widespread use of the automobile brought the city closer. Young people sought entertainment and socialization at dances, theaters, and schools. The Catholic churches, however, remained a dynamic force in the lives of farm families. Regular church attendance, feast days, and festivals were observed faithfully in Stearns County, and parish priests discouraged Catholic children from socializing with children of other faiths.[29]

Farm families continued to work with neighbors during haying and harvest times and in the construction of new farm buildings, and they helped one another in times of natural disasters such as tornadoes or snowstorms. There was less socializing than in earlier days, however, because larger dairy herds producing milk all year round and larger capital investments meant farmers had to spend more time at home. Morning and evening chores took longer. Visiting was done in the late evenings and on Sundays or church holidays.

So with a new standard of living brought about in part when Stearns County farmers responded to government encouragement toward mass production during World War II, farmers began to rely on government for price supports and subsidies. Much of the independence and freedom valued so highly by their immigrant parents and grandparents was relinquished. Government had become a permanent member of the dairy farm family.

UNCLE SAM JOINS THE FAMILY
Dairy Farms and the Government

Early subsistence Stearns County family farmers lived and conducted their business with little outside influence or control. With increased production, they came to depend more on outside markets and capital and less on family labor. And as problems beyond local control arose, they began to depend on the government to solve them. In the third phase of dairying in Stearns County, government became a permanent member of the family, and the family lost much of its former liberty in choosing how to live.

Early Government Involvement

Government and dairy farmers existed side by side from the time dairy farming began in the county. Most Stearns County settlers purchased their land from the state, from corporations, or from land speculators, but the Pre-emption Acts of 1841 and 1855 and the Homestead Act of 1862 made it possible for people with little money to obtain land for farming, too. Furthermore, although farmers relied on their own resources as a way of life, they accepted government aid after natural disasters such as blizzards, droughts, or grasshopper plagues, in the form of money, grain, and seed for planting as well as food, clothing, and shelter.[1]

Ag Schools

Other groundwork, too, had been laid by the federal government for the early dairy farmers in the county. The U. S. Department of Agriculture, established in 1862, obtained cabinet status in 1889. In 1862 the Morrill Act provided for the establishment in each state and territory of a land-grant college to teach agriculture and the mechanic arts. The Hatch Act of 1887 provided for the organization of an experiment station in each land-grant college offering four-year courses in dairying.[2]

Many Stearns County farm families took advantage of these agriculture courses by sending sons to study the latest methods of dairy farming. Ag schools taught farmers about new technology, about the adaptation of hardy alfalfa (hay that could survive the cold winters) in Minnesota, and about better modes of feeding, breeding, and housing cattle. Students took home their new knowledge to apply to family farming operations.[3]

The original College of Agriculture at the University of Minnesota, about 1870. The building was destroyed by fire in 1875. (Courtesy Minnesota Historical Society)

UNCLE SAM JOINS THE FAMILY

Daughters could attend summer ag courses including "lectures on home dairying, instruction and practical work in the manufacture of cheese especially adapted to home dairying in caring for milk and making it into butter after the most approved methods, and in testing cream and milk." Agriculture schools encouraged farm children to leave home for college while it bound them to the farm through newly acquired knowledge.[4]

The "Father of Cooperatives"

Theophilus L. Haecker came to Minnesota from Wisconsin in 1891 to start a school of dairying at the University of Minnesota College of Agriculture. Haecker believed that if farmers owned their own creameries together, they would be motivated to apply better methods in general operation of their dairying. On behalf of the University of Minnesota Ag School, Haecker traveled through Minnesota, expounding the virtues of cooperatives. By 1898 the state had 560 cooperative creameries and Haecker had come to be known as the "Father of Cooperatives," as well as the "Father of Dairying in Minnesota."[5]

But creameries had their government problems too: Antitrust laws prevented groups of creameries from marketing produce together for railroad carlot rates, and until the federal government intervened, larger corporations controlled the prices paid to farmers.[6]

The Federal Government and World War I

In World War I, the federal government made its first direct involvement in dairy farming by encouraging increased farm production. Dairy prices also rose with wartime demands at home and abroad. Stearns County farmers responded by expanding dairy herds and switching to winter dairying, producing more milk than ever before. By 1920 there were 42,543 dairy cows in Stearns County, fifty-seven to each square mile.[7]

Legislation between the Wars

After the war ended in 1918, prices dropped with demand, and agriculture leaders tried to unite cooperatives and other associations in support of a national farm policy. In 1922 a United States senator from Minnesota, Andrew Volstead, authored the Capper-Volstead Act, enabling farmers to jointly market their products without being prosecuted for antitrust violation. This allowed farmers

Theophilus Haecker, the "Father of Dairying in Minnesota." Haecker traveled extensively through Minnesota at the turn of the century expounding the virtues of cooperative creameries. (Courtesy Minnesota Historical Society)

Andrew Volstead, author of the Volstead Act of 1922. (Courtesy Minnesota Historical Society)

UNCLE SAM JOINS THE FAMILY

Empty milk cans ready to be picked up by farmers at the Paynesville railroad depot, 1916. Creameries utilized railroads to ship their milk products to outside markets. (Courtesy Paynesville Historical Society)

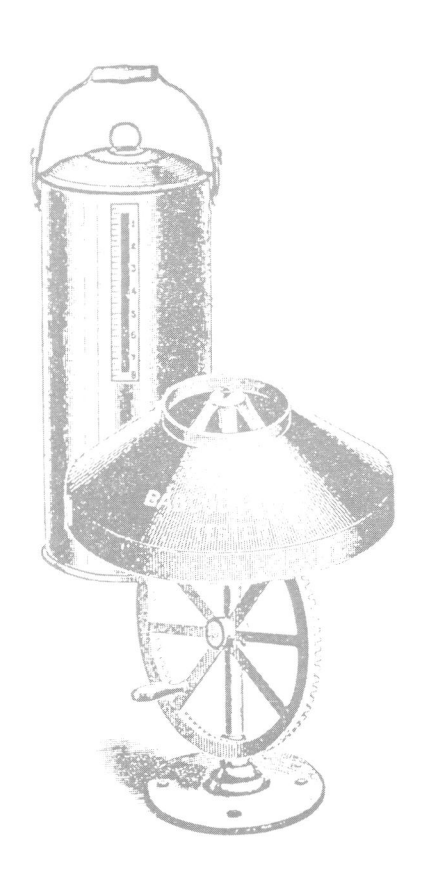

and creameries to combine carlots of butter for better shipping rates and realize more profit and bargaining power. At the same time, they relinquished some of their independence. Creameries had to conform to specific standards of production and sanitation set up groups of creameries to realize these profits. Farmers had to be more careful in sterilizing equipment and handling milk to receive the price differential paid for high-grade milk and cream which, in turn, yielded the best price for the creameries. The new regulations changed family roles, especially for children: "Kids couldn't do as much after government rules had to be obeyed," said Gerhard Gamradt. "Good, sweet milk brought a better price, so parents paid more attention to everything. Kids did more feeding and easier work, not as much with the milking."[8]

The government continued to respond to farmers' demands for controls and regulations. In 1929 the Federal Farm Marketing Board was established to secure orderly marketing and price stabilization for farm production. Although the board went out of existence in 1933, many agricultural historians believe this legislation was the first step toward permanent government involvement in farm policy.[9]

During the Great Depression of the 1930s, New Deal legislation gave emergency credit to low-income farmers, liberalized government lending agency programs, and united them under the Farm Security Administration. Attempts were made to obtain parity prices—a farm product price level maintained through government support to give farmers purchasing power equal to that in a given base period.[10]

Under the Agricultural Adjustment Act of 1933 farmers received payment for reduced acreage, and subsequent farm surplus policies were based on this principle. Most dairy farmers in Stearns County, however, were not directly affected by this act at the time because they raised crops for feeding cattle rather than for sale and few could afford to let fields lie fallow. The act, declared unconstitutional in 1936 (it was replaced by an act of the same name that incorporated parity pricing in 1938), nevertheless encouraged large farm operations, a step in the "bigger is better" approach to family farming.[11]

In 1937, cooperatives received help via the Agricultural Milk Marketing Agreement Act, designed to guarantee a stable market for milk. Agricultural historian Ingolf Voegler believes that today's regulatory framework for milk and dairy products is rooted in this law because it treats all farmers as equals. Actually, since not all farmers have equal access to resources such as land, capital, technology, and information, farmers like those on small Stearns County family dairy farms have not been treated on an equal basis.[12]

The Expanding Government Role after World War II

World War II marked a new era in major direct government involvement in dairy farming. Dairy farmers responded to government demands for increased production by producing over 50 percent more than they had during World War I. There were other forces at work, too: Young men left the farm for military service or higher-paying jobs in the cities, women joined the work force, and new dairy products were promoted—dry and powdered milk products for human and animal consumption, as well as condensed milk products and new brands of cheese. Oleomargarine was advertised as less expensive than butter. Farmers responded by purchasing machinery, increasing their herds and crop acreage, fighting oleomargarine proponents politically, planting new strains of hybrid corn and other crops, and applying pesticides, insecticides and commercial fertilizers. Agricultural products had become a government cartel by the end of World War II.[13]

Disease Control

As government became more involved in dairy production, animal disease control also became important. Mandatory testing of cattle became routine, but slaughter was the only method then of eradicating many diseases potentially harmful to humans and a herd of milking cows that took years to develop could be destroyed in one day. Federal and state governments partially compensated farmers who had to kill cattle with bovine tuberculosis or Bang's disease (contagious abortion). But veterinarian R. W. Page, Alexandria, said the compensation was never enough: "There's a big difference between a milk cow and a hamburger."[14]

Slaughtered herds took years to rebuild, and many families never rebuilt theirs. Stearns County was not exempt, and many families suffered great hardship when disease affected their herds. The Jerome Pfau family, Freeport, was required to destroy seventeen of its herd of forty-five cows. The Hansen family of Rockville had ten head of cattle condemned and "went out of dairying for a few years because of it."[15]

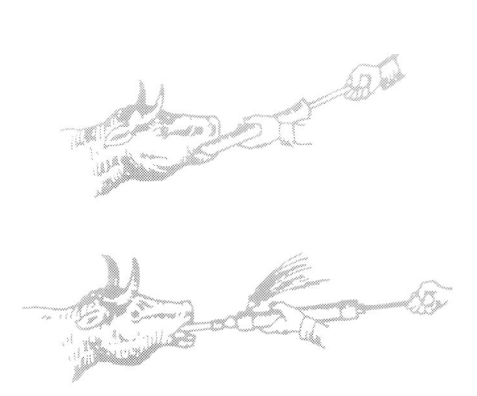

UNCLE SAM JOINS THE FAMILY

Most dairy farm families kept a bull for breeding purposes until the 1960s. Jerry Pfau, Freeport, posed with the Pfau family bull, a purebred Shorthorn, Duke of Glen Oak, in 1930.

According to St. Cloud veterinarian Ralph Ganz, "State boards of health worked with government to eradicate diseases. Fortunately, today there are preventions such as herd-health checks, calf-hood vaccination and sulfa drugs and penicillin." Artificial insemination reduced contamination through mechanical breeding methods and by eliminating the herd bull, which could transfer disease to the entire herd. Minnesota was declared free of bovine tuberculosis in March 1976 and of Bang's disease in October 1984.[16]

DHIA and the Extension Service

The Stearns County Extension Service worked closely with veterinarians, farmers, and government in eradicating livestock disease. It also helped establish county cow-testing organizations such as the Dairy Herd Improvement Association (DHIA), which developed an organized system of keeping dairy records related to health, herd production, reproduction, feeding, and identification of animals. Casimer Weller, a director on the Stearns County Dairy Herd Improvement Board said, "Good cattle knowledge is passed from father to son, but an unlimited amount of knowledge can be determined from records." DHIA records help farmers determine which cows to cull and which to feed for more production, and to evaluate cost and profit on an ongoing basis.[17]

Extension service agents also worked as educators and advisors for farm committees, farm clubs, cooperatives,

Stearns County Extension agents conducting a farmers' tour studying livestock about 1925. County agents played a major role in the gradual expansion of government involvement in the everyday life of dairy farmers.

UNCLE SAM JOINS THE FAMILY

A meeting of the St. Cloud Area Holstein Association directors in the 1940s: (standing) Sylvester Rademacher, Brother William Borgerding, Math Jennissen, Ed Fiedler, Earl Eichler, (seated) Bob Gates, and Herman Imdieke.

marketing organizations, homemakers' groups, and other rural activities, including 4-H. This and other young clubs organized after passage of the Smith-Lever (1914) and Smith-Hughes acts provided federal support to extension agents, to encourage active participation by farm youths in agriculture. Emphasis was placed on practical knowledge combined with leadership training in order to stimulate interest in farming as a vocation. The clubs enhanced social life, too. Dances, plays, field trips and other activities guided by parents, other adult leaders, and teen volunteers were encouraged. Families attended meetings and activities together, often in rural school or area churches. By 1922 every county in Minnesota had youth exhibits at the Minnesota State Fair, 1,454 people, aged seven through nineteen, were involved in 4-H in Stearns County in 1978.[18]

A 4-H cattle exhibit at the Stearns County Fair in July 1960. Four-H and other farm groups have traditionally encouraged children's involvement in dairy farming and promoted the ideal of the family farm.

31

UNCLE SAM JOINS THE FAMILY

The 1950s: Social Welfare

Stearns County farmers, perhaps because they held to many of the conservative German Catholic values and standards of their ancestors, lagged behind the nation in adapting to technology and change. By the mid-1950s, however, the "bigger is better" attitude encouraged by the government took root, and the average farm had increased from 167 acres in 1936 to 188 acres in 1956.[19]

In 1955 another government program greatly changed the lives of Stearns County farm families: Farmers became eligible for social security. Extension agents were called upon to work with hundreds of father-son partnerships and small farm corporations because social security made it possible for parents, upon reaching social security age, to retire but, because farm help was expensive and difficult to find, to remain on the farm as part of the work force.[20]

The Henry and Ceil Salzl farm near St. Martin in 1955. The farm typified Stearns County dairy operations in the 1950s and 1960s.

Despite government support, dairy farm families still worked long, hard hours. The Stearns County Extension report for 1958 said: "Most farm families have a pretty heavy livestock set-up and on such farms, in too many cases, the farm family is badly overworked. The income picture is such that it is next to impossible to pay the wage scale that is offered in industry for labor that might be needed on the farm . . . it is well to mention that the urban viewpoint seems to be that mechanization of farms

UNCLE SAM JOINS THE FAMILY

is making the job easier. That is probably true in a crop style of farming, but where the livestock program is carried out, that is hardly the case . . . Good management is entering the picture."[21]

The report added that the average age of farmers in the county was between fifty-five and seventy years, and "the changing status of rural [farm] people requires attention. What is meant is that there are more older people in the area now than what was true in years past. . . . The matter of enough income for these people to live happily and satisfactorily in this period of inflation is quite a problem."[22]

Joe Fabeck, St. Anthony, taking out milk left in the udder by hand after the electric milking, 1951. After the end of World War II, the average age of farmers rose dramatically as more and more young people left farms for urban jobs.

County extension workers began to work less with agricultural and horticultural problems and more with social ones. The 1961 annual report noted: "Modern advances in technology and the rapidly changing economy have had a strong impact on family living and have resulted in requests for new kinds of assistance from the Extension Service. Family economics, home management, buying, human relations, food and nutrition, clothing, housing, citizenship, health and safety, conservation, and problems of low incomes are some of the major areas in which Extension now conducts family living programs."[23]

The Demise of Creameries

While matters that Stearns County dairy farm families had traditionally resolved by themselves became issues of

A North American Creamery (Paynesville) bulk hauling truck, c. 1940. Large milk product companies, known as "centralizers," gradually replaced coop creameries as the main buyers of farmers' milk. (Courtesy Paynesville Historical Society).

public concern, the creameries that initially provided the wider dairy market upon which the new worldly life-style of farmers is based, were closing down. Once the pride of small farmers and their communities, the creameries could no longer maintain the volume required to compete with larger corporations achieving greater economies of scale.

In 1943 Stearns County was the largest farmer-owned cooperative butter-manufacturing area in the state, with forty-three cream stations, thirty-five creameries, and two cheese manufacturing plants; farmers worked with local creamery boards of directors to market their milk. By 1971 four major marketing organizations served the area, in 1979 only two Stearns County creameries still manufactured butter, and in 1985 there were none. By the 1960s bulk tank trucks picked up milk at the farms, and the only contact farmers had with their creameries was with truck drivers. Wilfred Schulte, a Stearns County butter maker, shared the belief of many farmers: "State and federal rules and regulations ruined small creameries."[24]

Government Price Supports

As dairy farm families relinquished personal contact with local cooperatives, their finances depended not on their own labor but on how much commercial marketing organizations could pay. The bottom line was money, and in order to make more, dairy farmers had to constantly upgrade their operations. Most early farm families paid cash for supplies, equipment, and improvements. Now farmers obtain loans, many from government lending organizations, and credit is very important. "Farmers were caught in the times," Gerald Buerman, Paynesville, recalled, and they have been subservient to government ever since.[25]

As farmers worked to produce more, to make more money, surpluses grew. The federal government paid farmers to let their fields lie fallow; it bought up surpluses for school lunch programs and for distribution to senior citizens and the poor. It provided low-interest loans to young farmers, paid subsidies, and every year attempted new measures to solve the problem of low income and high expenses.

At first the government attempted to assist farm families because farming was the nation's major industry. In 1940 farm residents made up more than half of all rural people, and the majority of lawmakers had rural roots. Today, however, only a small percentage of legislators is involved

UNCLE SAM JOINS THE FAMILY

in family farming, much less of the voting public has rural roots, and less than 15 percent of the total rural population lives on farms. Because few people in positions of power understand the problems of the small family farmers, few wish to save the way of life that the family farm represents.[26]

Farm laws have historically benefited large corporate farms and cooperatives that generate millions of dollars annually rather than small family farmers like those in Stearns County, where the average farm is 225 acres and farms are still owned and operated by farm families.

A Loss of Rural Identity

There are wide variations in the averages, and Stearns County is no exception. College-educated dairy farmers who have computerized operations and consultants on everything from feeding programs to home interior design have neighbors managing quite well with a herd of thirty cows in a stanchion barn with few technological advantages. Stearns County dairy farm families, however, have joined the mainstream of American life. A University of Minnesota Extension family life specialist was quoted in the *St. Cloud Daily Times* in 1980: "Today's farm families are less isolated, more worldly and more educated than 20 or 30 years ago. In fact, it's getting difficult to distinguish between farm families and their urban counterparts. The average farm family has two children, not eight. The birth rate has fallen faster for farm families that it has in the city during the last decade."[27]

The Resilience of the Family

Most Stearns County dairy farms are still family operations. Cold Spring dairy farmer Lydia Willenbring said: "The heavy work of years ago is less now, children's work is easier, but the whole family still works on most farms, and government programs should be based on supply and demand." Henry Banal, a veterinarian in the Sauk Centre area, said: "Farming is a business today, but the way of life is still part of it. Efforts should be made to keep the way of life part of farming." Banal said he always envied the dairy farm family life, that the presence of women and children, actively involved or only giving support, was apparent in the general appearance, in the attitudes of the men, and in the general operation of the farm. "Government doesn't take the family in farming seriously enough."[28]

Children feeding heifers on the Bechtold farm near Rockville, 1970. Dairy farming continues to be a family operation in Stearns County.

Becky and Richard Kotten, Farming, 1987. Children are still an important part of the labor force on Stearns County dairy farms.

UNCLE SAM JOINS THE FAMILY

The belief in the sanctity of the family farm is echoed by Casimer Weller, Belgrade: "Stearns County is livestock and family-farm conducive, not profitable as cropland. Our rural heritage is good; there is room to live and breathe. The good life on the family farm is still there. I recommend it for my children. Dairy farmers are good stewards of the soil, too. They are willing to save our land from erosion and wind." In regard to farmers losing their land in recent years, Weller said, "Eliminating poor farmers, poor land, poor cows, is all a continuing process of farming. It always has been"[29]

On the French family farm, Brooten, as on most Stearns County dairy farms, the entire family helps with the milking chores. Pictured are Bob, Brian, and Ron in 1984.

Younger dairy farmers like Robert and Audrey French, Brooten, agree that "Dairy farming is still a way of life. The whole family helps with the work. That's the only way you can make it, along with good management." The Frenches say they have continued many traditions of their parents like preserving home-grown fruits and garden vegetables, but "It's not possible for government to get out of farming anymore, so you make adjustments." Robert and Sandy Hemmesch, Paynesville, say, "We like the life ourselves. We hope our children will continue on the farm. We start them helping with the work while they're young."[30]

Familes still operated 1,842 dairy farms in Stearns County, with 82,700 milk cows in 1986. The average DHIA production per cow was 15,984 pounds of milk annually; the overall average production was 12,000 pounds.[31]

In 1986 there was a certain nostalgia in Stearns County for the farm family life of the past. Idealists held on to the

UNCLE SAM JOINS THE FAMILY

concept of family dairy farming as a way of life. Realists treated family farming as businesses operated with family labor. Both felt family dairy farming was a good life, that it offered some choice along with good clean country living. On good days all would agree with a Paynesville short-haul trucker who said, "Stearns County dairy farmers are the best anywhere," and almost all want their children to continue farming. As one retired couple living on a family dairy farm near Luxemburg said, "We have always liked it here. We have no desire to leave. Why should we?"[32]

This is an ideal young milking U. S. Holstein cow as determined by the Holstein Friesian Association of America. By the 1960s, the breed was predominant in Stearns County because of the large volume of milk it produced.

NOTES

Full publication information is included in footnotes in lieu of a bibliography. A comprehensive bibliography is available at the Stearns County Historical Society.

ONE OR TWO FAMILY COWS

1. Wendell Berry, *The Unsettling of America, Culture and Agriculture* (New York: Avon Books, 1977), 44; Jonathan Norton Leonard, *The First Farmers* (New York: Time-Life, Inc., 1973, 86-99; Lyman Carter, *The Beginnings of Agriculture in America* (New York: McGraw-Hill Co., 1923), 256.
2. Martin Lawrence, *The Homesteader's Handbook* (New York: Mayflower Books, Inc., 1979), 163; for breeds of cattle see Ralph W. Wayne, *A Century of Dairying and Registered Holsteins* (Detroit Lakes, Minnesota: Lakes Publishing Co., 1976), 57; *Guernsey Breeders Journal: Commemorative Issue* (July 1977), 47-48; Alvin H. Sanders, *Shorthorn Cattle* (New York: Sanders Publishing Co., 1918), 54; A. D. Allen, ed., "Oxen," *American Agriculturist*, vol. 7, (1848), 339.
3. Ron Hamel, "The Growth of Dairy Farming in Early America," *Hoard's Dairyman: Bicentennial Issue*, (November 1975), 791-804; Paul W. Gates, *The Farmer's Age: Agriculture 1815-1860* (New York: Holt-Rinehart and Winston, 1960), 232-248.
4. Herman Kohorst, interview with author, 1979, author's private collection; William Bell Mitchell, *The History of Stearns County Minnesota* (Chicago: H. G. Cooper Jr., and Co., 1915), 1007.
5. Ibid., 830, 1049.
6. Ibid., 926.
7. Arnold Borsheim, "The Nielson Sage," recorded, 12 December 1957, Family History Files, Stearns County Historical Society (SCHS).
8. August Lemke, interviewed 5 May 1937, WPA Files, SCHS; Merrill and Joann Grohman, *The Cow Economy* (Dixfield, Massachusetts: Coburn Farm Press, 1975), 41.
9. Bertha Carpenter, interview with author, 2 February 1987, SCHS; Adaline Koshiol, interview with author, 27 March 1987, SCHS; Merrill E. Jarchow, *The Earth Brought Forth* (St. Paul: Minnesota Historical Society, 1949), 208.
10. E. L. D. Seymour, ed., "Dairy Machinery for the Farmer," in *Farm Knowledge* (New York: Doubleday, Page & Co., 1918), 135; Michael Partridge, *Farm Tools Through the Ages* (Boston: Promontory Press, 1973), 218.
11. Math Ohnsorg, interview with author, 11 March 1987, SCHS; Laino Alato, "Life in a Log Cabin," in *The Gopher Reader* (St. Paul: Minnesota Historical Society and Minnesota Statehood Centennial Commissioln, 1966), 92.
12. Arlene Willenbring, interview with author, 4 February 1987, author's private collection (Willenbring donated her mother's butter print to SCHS); Henry Behnen, *History and Lifestyle of My Early Years* (Greenwald, Minnesota: author's private printing, 1981), 22, SCHS.
13. Koshiol added that some women in the Luxemburg area made all their cream into butter for local sale long after the creamery closed, well into the 1960s.
14. Amos Zum Brunnen, interview with author, 5 December 1985, author's private collection. Although Stearns County farmers did not generally make ripened cheeses, the Zum Brunnen family of Swiss heritage, Wright County, established a Limburger factory in Hasty in the 1920s.
15. Rosalia Fuchs, interview with author, 20 September 1984, author's private collection; Alfred Ebnet, interview with author, 25 March 1987, SCHS.
16. Grover Dean Turnbow, *The Ice Cream Industry* (New York: John Wiley & Sons, Inc., 1928), 1-3.
17. Hedwig Winter, interview with author, 27 June 1987, author's private collection. "Sand land" was considered poor crop land; during very dry years almost no crops could be raised there.
18. "George Kulzer, 1831-1912: A Continuing Story of a Stearns County Pioneer," *The Albany Enterprise*, 15 June 1976 (from the diary of George Kulzer translated by daughter-in-law Mary Kulzer in 1935, completed by granddaughter Ramona Kulzer in 1970, and serialized in *The Albany Enterprise*, SCHS).
19. Clara Symanietz, interviews with author, 1980-1982, SCHS and author's private collection; Marilyn Salzl Brinkman and William Towner Morgan, *Light From the Hearth* (St. Cloud, Minnesota: North Star Press, 1982), 99-103.
20. Clara Jenc, interview with author, 16 March 1987, SCHS.
21. Bill Vouk, interview with author, 23 April 1987, SCHS.
22. Bertha Zniewski, interview with author, 8 November 1982, author's private collection (Zniewski, president of the Paynesville Historical Society, said patterns for the shoes are still in the possession of Mrs. Ernestine Manz, Paynesville.) Alexander Korte, "The Korte Family," unpublished family history recorded, 20 May 1964, SCHS.
23. Symanietz.
24. Mitchell, 974; Kulzer.
25. Ebnet. He also said more was expected of oldest children. As the oldest in his family, Ebnet spent three years in seventh grade before quitting school because he missed too many days while helping out on the farm.
26. Brinkman and Morgan, 74.
27. Ibid., 73-119.

NOTES

THE TURN TO DAIRYING

1. Sigfried Giedion, *Mechanization Takes Command* (New York: Oxford University Press, 1984), 131; G. E. Fussel, *The Farmer's Tools* (London: The Mayflower Press, 1952), 209; Stearns County Assessment Rolls, 1890, Albany, Minnesota Historical Society.
2. "Industry Trebles During Period from 1914 to 1926," *St. Cloud Daily Times*, 8 June 1926, 1; Edward Weist, *The Butter Industry in the United States* (New York: AMS Press, 1968), 21.
3. Mitchell, 1419-20.
4. John Schwinghamer, interview by Walter Haupt, 17 November 1937, WPA Files, SCHS; "Stearns County Dairy Industry," supplement of *St. Cloud Daily Times*, 21 June 1925, 5.
5. "Stearns County Dairy Industry," 51.
6. Ibid.
7. Art Borgmann, interview with author, 27 January 1987, SCHS; Bill Vouk, interview with author, 24 April 1987, SCHS; Bill Vouk, interview with author, 24 April 1987, SCHS; William Cooper, interview with author, 31 March 1987, SCHS.
8. Fussel, 209; Jerome Pfau, interview with author, 22 January 1987, SCHS; Gerhard Gamradt, interview with author, 4 February 1987, SHCS.
9. Jenc.
10. Ceil Salzl, interview with author, 1984, author's private collection.
11. *A Historical Survey of American Agriculture Yearbook #1783* (Washington D.C.: U. S. Department of Agriculture, 1941), 288; *1923 Stearns County Extension Service Annual Report*, 3-6.
12. "Stearns County Dairy Industry," 5; for pictures of barns designed by county extension agents, see Stearns County Extension Service annual reports 1914-1925, SCHS; George Moonen, interview with author, 15 March 1983, author's private collection. Moonen said he played his concertina and accordion at such barn dances.
13. *Electric Helpers for the Farm Family* (Schenectady: General Electric Rural Electrification Section, n.d.), n.p.; for history of rural electricity in Stearns County see also Marilyn Salzl Brinkman, *Current and Kilowatts* (Melrose, Minnesota: Stearns Cooperative Electric Association, 1987).
14. Ibid.
15. Ibid; Winnifred Claude, interview with author, 31 August 1986, Stearns Cooperative Electric Association Files.
16. Robert Imdieke, interview with author, 6 August 1986; Stearns Cooperative Association Files.
17. Ebnet.
18. Henry and Ceil Salzl, interview with author, 29 April 1984, author's private collection; *St. Cloud Daily Times*, 15 August 1942, 6.
19. "Stearns 1935 Butter Value $3,065,299," *St. Cloud Daily Times*, 8 August 1936, 1; "National Defense," *1941 Stearns County Extension Service Annual Report*, 25, SCHS.
20. John T. Schlebecker, *Whereby We Thrive* (Ames: Iowa State University Press, 1975), 214-15.
21. "The Third Twenty-five Years, 1935-1960," *Hoard's Dairyman Centennial*, July 1985, 803-10.
22. "Dairy," *1941 Stearns County Extension Service Annual Report*, 25-27, SCHS; "Produce More to Win the War," *St. Cloud Daily Times*, 21 June 1943, 13-23.
23. *Hoard's Dairyman*, 11 June 1943, n.p.
24. Kenneth D. Ruble, *Men to Remember*, (Chicago: Lakeside Press, 1947), 256; "New Buttermilk Factory Built at Litchfield," *The Waverly Star and Tribune*, 31 March 1927, 1; "Albany Dedicates Dry Milk Plant Tuesday," *St. Cloud Daily Times*, 15 June 1936, 11 (Since "skim" suggested milk fed to animals, manufacturers soon tagged the new product "dry milk solids" or "nonfat dry milk solids."); "Kraft Plays Major Role in History of Area," *Melrose Beacon*, 20 May 1987, 15; "The Third Twenty-five Years," 803-10.
25. U.S. Department of Agriculture, Economics, Statistics, and Cooperative Service, Structure Issues of American Agriculture: Agricultural Economic Report 438 (Washington D.C.: Government Printing Office, 1979), 75.
26. Ibid.
27. "Livestock, "*1953 Stearns County Extension Service Annual Report*, 6, SCHS.
28. "Farm Management," *1953 Stearns County Extension Service Annual Report*, 30; "Introduction," *1952 Stearns County Extension Service Report*, 3; and "Introduction," *1955 Stearns County Extension Service Annual Report*, 6, SCHS. In 1978 there were 70,000 cows in Stearns County.

NOTES

UNCLE SAM JOINS THE FAMILY

1. Fred A. Shannon, *The Farmer's Last Frontier: Agriculture, 1860-1897* (New York: Fred A. Shannon, 1945), 255,57-60.
2. Theodore C. Blegen, *Minnesota: A History of the State* (St. Paul: University of Minnesota Press, 1975), 420-3.
3. Ralph E. Miller, *The History of the School of Agriculture, 1851-1960* (St. Paul: University of Minnesota Institute of Agriculture, Forestry, and Home Economics, 1960), 27. More than ten thousand Minnesotans were students during its seventy-two years of operation as a department.
4. *Minnesota Farmers' Institute Annual Number 9, 1896* (Minneapolis: Tribune Printing Co., 1896), 29.
5. Kenneth D. Ruble, "Father of Cooperative Creameries," *Land O' Lakes Mirror,* (December 1975), 14; also Ruble, *Men to Remember,* 12; Mitchell, 1419.
6. Bill Vouk.
7. "Stearns County Diary Industry," 5; Joseph Schafer, *The Social History of American Agriculture* (New York: The MacMillan Co., 1936), 222.
8. Carol James, *Minnesota Cooperatives Buy Volstead House,* pamphlet (Washington D. C.: U. S. Department of Agriculture, June 1977), 11; Ruble, *Men to Remember,* 47; Gamradt.
9. Schafer, 281.
10. "Farm Home Administration," *Encyclopedia Americana,* Vol. 11, (1967), 25-26.
11. "Agriculture Adjustment Act," *Encyclopedia Americana,* 1, (1967), 342.
12. Ingolf Voegler, *The Myth of the Family Farm: Agribusiness Dominance of United States Agriculture* (Boulder, Colorado: Westview Press, 1981), 294: U. S. Department of Commerce Bureau of Census, "1974 Census of Agriculture, Stearns County Minnesota, July 1976; Voegler offers insights into why the American family farm is disappearing; see also Raymond T. Howard and William M. Smith, eds., *The Family in Rural Society* (Boulder, Colorado: Westview Press, 1981); Don A. Dillman and Daryl S. Hobbs, eds., *Rural Society: Issues for the 1980s* (Boulder, Colorado: Westview Press, 1982).
13. "The Fight on Yellow Oleo," St. Cloud Daily Times, 17 January 1949, 10; see also S. F. Piepma, *The Story of Margarine* (Washington D. C.: Public Affairs Press, 1970); and Schlebecker, 214-15.
14. R. W. Page, interview with author, 29 June 1987; author's private collection; Ralph Ganz, interview with author, 7 May 1987, SCHS.
15. Pierre Hansen, interview with author, 19 January 1987, SCHS; Jerome Pfau.
16. Ganz; Schlebecker, 268. Plaques were presented to the Minnesota Association of Veterinarian Medicine when Minnesota was declared bovine tuberculosis and bang's free.
17. *Stearns County DHIA Annual Meeting Report* (Sauk Centre: DHIA Laboratory, 1986), 3. See also "Official DHIA AM/PM Testing," *Minnesota DHIA Bulletin,* (St. Paul: University of Minnesota Agriculture Service and USDA, 1986); Casimer Weller, interview with author, 1 April 1987, SCHS.
18. Marianne Kibler, *4-H in Stearns County,* pamphlet (Stearns County Extension Service, March 1978), SCHS; see also John Ritter, "4-H Grew From Agriculture Clubs of Early 1900s," *The Farmer,* 100 Years, 1882-1982, 100 (1 May 1982): 62-65; author's personal experience with 4-H work and organization in Stearns County.
19. "Livestock," *1956 Stearns County Extension Service Annual Report,* 1; "Farm Management," 30, SCHS.
20. "Livestock," *1958 Stearns County Extension Service Annual Report,* 2, SCHS.
21. Ibid.
22. Ibid.
23. "Family Living," *1961 Stearns County Extension Service Annual Report:* "Introduction," SCHS.
24; "Minnesota Dairy Plants," (St. Paul: Minnesota Department of Agriculture, Dairy Industries Division, 1979), n.p; see also Bill Brady, "District I," Land O' Lakes, Inc., Report Number A422, 24 October 1986; Wilfred Schulte, interview with author, 4 March 1987, SCHS.
25. "Dairy," *Facts About Minnesota Agriculture* (St. Cloud: Central Minnesota Agriculture Committee and St. Cloud Chamber of Commerce, 1971), 2; Gerald and Mary Jo Buerman, interview with author, 5 March 1987, SCHS. Although many Stearns County farm families objected to government involvement in their farming operations, during National Farmers' Organization withholding actions in 1970s, a number of individual farmers took part in the demonstrations, but many were against the actions and they did not affect Stearns County dairy farmers in general.
26. U. S. Department of Agriculture, Economics, Statistics, and Cooperative Services, *Structure Issues of American Agriculture,* 284. Rural has come to mean any town or city with fewer than 10,000 residents.
27. Ron Pitzer, "Rural Families Undergo Change," *St. Cloud Daily Times,* Family Farm Section, 1 March 1980, 22E. Although dedicated to Stearns County farm families, the fifty-six-page section included only two articles and pages about women and children.
28. Jenc; Lydia Willenbring, interview with author, 29 January 1987, SCHS; Henry Banal, interview with author, 16 March 1987, SCHS.
29. Weller.
30. Robert and Audrey French, interview with author, 1 April 1987, SCHS; Robert and Sandy Hemmesch, interview with author, 5 March 1987, SCHS.
31. Francis Januschka, personal correspondence, 29 May 1987. Although this study concentrates on dairy farm families in Stearns County, of 3,450 farms in the county, only 1,842 are dairy farms. The others are beef, turkey, hog operations, or hobby farms that produce over $1,000 worth of farm products annually; some are crop farms.
32. Koshiol; Norman and Genora Bork, interview with author, 12 March 1987, SCHS.

BRINGING HISTORY HOME

Annette Atkins

Annette Atkins is particularly interested in the role of the family in the realm of American social history. Her most recent work, *Harvest of Grief: Grasshopper Plagues and Public Assistance in Minnesota 1873-78* was published in 1984 by the Minnesota Historical Society. She is an Associate Professor of History at St. John's University, Collegeville, Minnesota.

Bringing Home the Cows begins not with an exhibit panel or essay but in the mind's eye, in the imagination, and in the understanding that each of us is an actor in an important historical story.

We know almost intuitively that presidents and governors are important, as are great scientists, inventors, writers, philosophers, and musicians. Often in learning history, we study these great people—people who made major decisions, who led us into or out of wars, who made wise or unwise decisions for their nations, who behaved nobly or reprehensibly, or whose inventions, paintings, ideas, or books changed the world. And we recognize single great events: the signing of the Declaration of Independence, the bombing of Hiroshima, the assassination of Martin Luther King, Jr., the explosion of the *Challenger*. We know these people and events should be noted.

But most of the people who lived in the past were not "great men"; most weren't great and half weren't men. Only a very few played large roles in the grand events of their time. Most were men and women, much like us, who lived private lives in fairly circumscribed locales. *Bringing Home the Cows* is about such people.

Men and women on dairy farms make foreign policy only indirectly, as voters. They make war or peace only as soldiers, nurses, or involved citizens. They don't make their life's work of painting great pictures or writing great poetry. Nonetheless their lives are historically significant. Farmers—men and women—feed people. They care for animals. They create a way of life that integrates work and home, family and neighbors. As individuals these people matter deeply to their families, friends, and communities. As a group they matter to the state and to the nation.

As historical actors they can make a difference to each of us, too. They can teach us how to live—by comparision or contrast. They tell us that our struggles and advances are not ours alone, that our families did not invent trouble or success. And by showing us that in other times and places people made decisions quite different from our own, they help us expand the range of our own possibilities.

To take the fullest meaning possible from the story, we might think first about history as a way of seeing the world rather than as a collection of dates, names, and facts. History organizes information around time and looks at relationships between past and present. It calls us to ask: How is the present like and unlike the past? What factors in their time and place led them to lead their lives as they did, to make the decisions they did? The more information we gain, the more questions we might ask: Were they happy? Why did they stay married? How did they live without "modern conveniences"? If we keep our ears and minds open we can hear their answers and sometimes be surprised.

Second, we can try to see people in the past as neither more nor less human than us. Dairy farmers in Stearns County were neither bigger nor smaller than we are; their emotions, imagination, hopes, fears, and motives have been no less complicated than our own. If we romanticize their ability to work, withstand trouble, or smooth out family relationships, we become blinded by our own inadequacies. Or

BRINGING HISTORY HOME

if we assume them to have been simple rustic people who didn't know any better, we miss their wisdom. We shouldn't fall for either the good-old-days or the bad-old-days syndrome.

History as such does not repeat itself. Few things happen exactly as they've happened before. What does repeat is the structure of lives, the fact that there are issues and problems to be faced, decisions to be made, definitions to be created, that there is a need to work out decent and honorable lives. Public issues may change over time, but personal issues — marriage, raising a family, making a living, the desire to belong — tend to say the same.

Bringing Home the Cows covers more than a century of Stearns County history played in a larger national setting. The people of the county knew about, participated in, and were affected by national events; as white settlers arrived here in the 1840s and 1850s, the United States faced several enormous and complicated issues: slavery, reform, westward expansion, sectionalism, the relationships of whites and Indians. Whatever your politics, as a farmer in Stearns County in 1860, you would have felt the fear of an impending civil war. You might have worried about a brother, husband, or son being drafted, about signing up yourself, about what would happen to your farm, your neighbors, and community. You would certainly have been in less immediate contact with daily happenings in Washington, D.C., than today, but like many in the nineteenth century, you would probably have been well informed. Men knew what issues were at stake when they cast their ballots in local, state, and national elections (women did not get the vote until 1920). These were not simpler times than our own.

As a farmer in the 1870s and 1890s you would have worried about and been affected by economic depressions as devastating then to farm families as more recent economic events. Weather and natural disasters posed grave threats. Candidates of all parties paid attention to farm questions, and the People's (Populist) Party was founded almost exclusively because of feelings that the American government was not doing enough to support the family farm.

In 1920, the United States Bureau of Census reported for the first time that a majority of Americans lived in urban places. The nation seemed to turn away from farming and farmers when sophistication, glitter, and wealth replaced perseverance and work as American values and F. Scott Fitzgerald's Great Gatsby replaced the "noble yeoman" as the ideal to which many Americans aspired. The 1920s, recalled as "roaring," are characterized by flappers, the Charleston, and bootleg liquor.

Stearns County had its share of bootleg liquor, but in other ways it diverged from the American pattern. The majority of Minnesotans did not live in urban places until 1950 and although a majority of Stearns County citizens lives off the farm, the values are probably still more rural than urban. In Garrison Keillor's fictional Lake Wobegon, simple rural people are valued more than sophisticates. This holds true for the real Stearns County.

In other ways, too, Stearns County farmers adhered to different ways. Despite harsh American reaction to all things German during World War I, German was spoken and honored in many parts of the county. It is difficult to uphold values apart from the larger world, and the county suffered from the prejudice of outsiders, but German intonation can still be heard in the speech of many. And as the nation rediscovers the value of ethnic roots, the strong heritage of the county is increasingly respected.

BRINGING HISTORY HOME

If the story of Stearns County dairy farmers begins in ourselves, it also ends in self-reflection. The images and artifacts, what historians call "material culture," recreate a way of life and show how it has changed over time. Underlying the story are questions that apply broadly and invite us to think about the role of technology, for instance, in our lives: What has been the impact of television on our families and communities? Has it improved our lives? Has it weakened family ties, increased aspirations, decreased satisfaction, made us hungrier, more acquisitive? Has it informed us better, made us more worldly, more tolerant, more or less provincial? What about cars, radios, blenders, microwave ovens, computers?

Bringing Home the Cows does not answer such questions; rather it suggests that changes are not simple "facts" but decisions that have implications and consequences in people's lives. Discovering the kinds of change, asking the questions, and finding the implications is important historical work for each of us to do on our own.

NEITHER HOUSEWORK NOR WAGE LABOR: WOMEN'S DAIRY WORK

Martha Blauvelt

Martha Blauvelt specializes in the history of women in America. She has written articles on women and revivalism for Volumes I and II of *Women and Religion in America*. She is an Associate Professor of History at the College of St. Benedict, St. Joseph, Minnesota.

Although until the late nineteenth century the majority of Americans lived on farms, today we know curiously little about their work: We suspect it was often tedious and backbreaking, but few historians have charted its evolution, nature, and meaning in sufficient detail. This is especially true of women's work. Neither housework nor wage labor, it does not fit into the usual historical categories. As a result, we know relatively little of how women's farmwork changed over time, or of how it has varied on different types of farms. Still, women's work was absolutely essential to the farm economy, contributing to family prosperity, and determining women's status in a rural culture.

Dairy farming provides a particular interesting example of gender roles on the farm and in agricultural economy. *Bringing Home the Cows* provides glimpses of women's work on nineteenth- and twentieth-century dairy farms. In the first stages of settlement, the mostly German Catholic pioneers of this area typically had one or two cows. It was women's work to milk and care for them, chores which provided dairy products not only for family use but for the market as well. Women's butter making in particular became essential for the purchase of family necessities. As farms moved beyond the settlement stage and concentrated on dairying, women increased their butter making and, with other family members, cared for larger herds of cows — this in addition to their housework and childcare. Women took special pride in their butter and marked it with a special imprint which proclaimed it as their product. Dairy work thus provided women with a source of pride and reputation and endowed women's work with greater visibility than might be the case on a grain farm or cattle ranch.

But as dairy farms expanded and emphasized milk production, technology changed the character and meaning of women's dairy work. Milk machines did not reduce the amount of women's work — women gained the task of cleaning them — but the new equipment belonged to men, much as the churn had belonged to women. Most significantly, the women's role in the production of butter for market also declined: creameries began to take over butter manufacture, and women lost this source of pride, local fame, and commercial participation. By the mid-twentieth century, women's work on the dairy farm had lost much of its visibility, although it had by no means diminished in amount.

How typical was this pattern? It is dangerous to generalize on the basis of the single — and perhaps singular — example of Stearns County, but a recent analysis of women's dairy work in southeastern Pennsylvania provides additional comparative material. Joan M. Jensen's *Loosening the Bonds: Mid-Atlantic Farm Women, 1750-1850* (New Haven: Yale University Press, 1986) suggests that women's agricultural labor can intersect with culture and affect their status in strikingly different ways.

Although Jensen studies English Quakers in Pennsylvania rather than German Catholics in Minnesota and focuses on the century preceding white settlement in Stearns County, many common patterns prevailed: in each region one ethnic and religious group dominated rural culture. Both German Catholics and English Quakers began as subsistence farmers and became commercial dairy farmers. This move into the market economy was based on a division of labor along gender lines. In both areas women making butter was essential to initial farm success while providing them a source of pride and public reputation. The fact that husbands made the imprints with which their wives proudly stamped their butter suggests the egalitarian implications of women's roles in dairy farming in both Minnesota and Pennsylvania.

There the similarities end. The Quaker women of southeastern Pennsylvania differed from their Minnesota counterparts in three ways: the role of technology in their lives, the scale of participation in the market economy, and the degree of independence and public activity their culture allowed.

Jensen and Brinkman show the very different roles technology can play in rural women's lives. Jensen found that before the eighteenth century English and American men and women worked side by side in the fields, harvesting grain with the sickle. Both sexes could easily wield that tool, but when the heavier scythe appeared and a cradle was added, the implements became gender-specific. Technological changes encouraged women to shift their work to textile processing and butter making, leaving fields to the stronger men. This entailed a seasonal change in work activity; women concentrated their work in the spring and summer—the prime churning period—instead of during harvest. This enabled farm girls to go to school, especially from January to April.

In early America, technological change cleared the way for female education and economic significance. In Stearns County, however, milk machines and creameries reduced women's public economic role. Many women expressed reluctance to turn butter making over to creameries, although they were powerless to halt this change.

Jensen copiously illustrates how economically significant women's dairy work could be in Pennsylvania. By the late eighteenth century, farm women in the "butter belt" around Philadelphia supplied not only all of the city's butter but sold to the coastal trade and the West Indies as well. The amount of butter they churned grew steadily into the 1840s. Commercial dairies (manufacturing 8,000 pounds of butter or more a year) produced only 2 percent of this butter; small farms making 200 to 599 pounds a year provided almost half of the butter, while middle to large-sized dairies produced the rest. The men cared for farm animals and raised crops, but women did most of the milking and all of the churning. They also packed the butter, took it to market, and sold it. The cash or goods obtained through butter sales fueled both the "consumer revolution" of the late eighteenth century and the market economy of the early nineteenth century. In Jensen's words, the churn symbolized "not the domestic arts of housewifery but the commercial arts of women alert to the demands of the market."

With no access to a market as large as Philadelphia or a coastal trade, Stearns County women never produced butter on a scale comparable to their early American sisters. While their smaller production brought in money and goods essential to their families, Stearns County dairywomen did not occupy so central or public an economic position. This makes their twentieth-century retreat from butter making no less poignant, for as German Catholic women, they had few other opportunities for public notice. Barred from the priesthood or even the possibility of being altar servers, they played no major public roles in the Catholic Church. Although sodalities, rosary societies, and Christian mothers' societies might form an important part of their personal or religious lives, such organizations had little public influence. And as newly enfranchised citizens, Stearns County farm women in the 1920s and 1930s often found their husbands telling them how to vote.

For the Quaker women of early Pennsylvania, central economic function was only one of several public roles they enjoyed. The Society of Friends allowed both women and men to be "Public Friends" or ministers. Although Quaker women often hesitated to become Public Friends, fearing they were too forward or might neglect family duties, by the late eighteenth century female ministers were very common in this sect. Many of the Quaker women of the butter belt were also strong abolitionists. Although they lacked the vote, they asserted their political presence through signing antislavery petitions, joining abolitionist societies and feeding, clothing, and hiding runaway slaves—politically subversive and illegal acts.

Ironically, although the Quaker dairy women of early Pennsylvania play public roles in several spheres and made substantial contributions to the rural economy, they lacked legal control of their earnings. Under common law, men and women became one at marriage, that one being the husband. Unless careful provisions were made in separate trusts — an unusual procedure subject to male approval — husbands owned any money their wives earned. Although Quaker women lived in a culture supporting female independence and public roles for women, they lacked a legal basis. Stearns County dairywomen, on the other hand, lived in a culture and economy that until recently offered few opportunities for self-assertion or public roles, while enjoying legal rights to the fruits of their labor that an early nineteenth-century feminist could only dream about. Still, culture did not encourage women to assert the equality implicit in their legal position.

The story of women's dairy work is one of hard work in both Pennsylvania and Minnesota. *Bringing Home the Cows* recognizes how essential that work was to the dairy farms of Stearns County. Through it we can imagine early rising, the exertion of washing pails and milk machines, pride in producing sweet butter, and affection for farm animals. But the meaning of dairy work was vastly different for women in the two states. For Quaker women butter making was only one of many public activities typifying their assertiveness in politics and religion. For Stearns County women, dairy work was an activity that occasionally pulled them into the marketplace but more commonly expressed their efforts to aid the family unit, a helping role they also played in church and community. Dairy farming exemplifies the importance of women's work, but there two cases suggest the vastly different ways gender and culture can intersect to determine women's status.

DAIRY FARM ARCHITECTURE IN STEARNS COUNTY

Fred Peterson

Fred Peterson has written and lectured extensively on the subject of rural architecture in Minnesota and the Upper Midwest. His recent works included articles in farmhouse architecture in *North Dakota History*, *Architecture Minnesota*, and *Common Places: Readings in American Vernacular Architecture*. He is currently Professor of Art History at the University of Minnesota at Morris.

*T*he buildings that farmers and businessmen have constructed to serve dairy farming in Stearns County have given a special quality to the landscape of the area. The characteristic group of structures that identifies the family dairy farm operation is present from the rolling hills of the eastern and central sections to the open prairie stretches of the northern and western portions of the county: a large barn and tall silo, one or more grain storage units, some smaller outbuildings, and a farmhouse. As portrayed in the picture "Her Majesty, the Dairy Cow," these farm buildings and all the members of the farm family that works the farm "serve" the dairy cow. In fact, "Her Majesty" has caused other kinds of structures and a variety of persons beyond the family farm to serve her. From local creameries, hardware stores selling dairy tools and machines, and implement sale yards to county and state fair pavilions, schools of agriculture, and farm experimental stations, the dairy cow has directly shaped the way many live and work, as well as formed the landscape of countryside, towns, and cities. One cannot overlook the Dairy Queen drive-ins and numerous ice-cream stands that delight us with the regal role that the dairy cow plays in our lives.

This picture appeared on the cover of a St. Cloud Times *special edition on dairy farming in Stearns County. It portrays the ideal dairy farm of 1923.*

Beginning in the early 1850s, the log buildings of the earliest settlers in the county set the pattern for later farms. The Michael Reisinger farm near Collegeville, as it was photographed about 1890, illustrates a grouping of structures around the farm cabin that either sheltered the cattle or protected the feed that was necessary to keep them through the long winter. These structures were small and simple, made of materials immediately available, and built according to a log construction method with which most German immigrants were familiar and that Yankee farmers from the East soon learned. Such farmsites were adequate for the subsistence level of farming.

After the end of the Civil War in 1865, a new wave of immigrants

DAIRY FARM ARCHITECTURE

The Reisinger Farm near Collegeville, Minnesota, around 1890. The simple log architecture was typical of Stearns County farms in the late nineteenth century.

The Hersing Family on their farm near Brooten, Minnesota, in 1915. This simple style of house easily allowed structural additions to accommodate growing farm families.

DAIRY FARM ARCHITECTURE

from Germany, Scandinavia, and Slovenia settled the central portion of the county along the Sauk River. By this time, milled lumber was available and the balloon-frame method of construction became the means to efficiently and economically build a small farmhouse as well as to construct the larger-scale buildings necessary for a successful farming operation. In most cases the barn and other outbuildings of the farm were realized first, according to the adage that "The house doesn't pay for the barn." The George Hersing farm near Brooten is typical of the early small frame farmhouse that sheltered the growing family. The Bellmont farmhouse in Rockford Township probably began as the small kitchen section on the far right of the structure. It is evident why the house grew. The one-and-one-half-story to two-story farmhouse built in an ell or T plan was the expedient pattern in which to construct "add-on" to an original structure as needed and/or desired. The ell or T plan type of Farmhouses that resulted from these stages of building became the most popular in the Upper Midwest during the 1860s and and 1870s.

The Bellmont family and their farmhouse in Rockville Township, Minnesota, in 1915. The house grew along with the family. The original section of the house is probably on the right side of the structure.

The kind of farmhouse that more specifically characterizes Stearns County farms was built during a period from about the mid-1880s to about 1920, when the entire county was settled and most dairy farms were well established. The Blonigan farmhouse near St. Martin is of this type. It is the square plan, two-story house with a pyramidal roof. Most of these large and impressive farmhouses in the county were built as frame structures but many were made of brick. Some have one or more dormers on the roof or a one- or two-story kitchen wing attached to the back of the house. The structures communicate the substantial success that farm families achieved after decades of developing a herd of dairy cows and working the land in such a way that its fertility and richness would be conserved. This type of farmhouse was made popular through mail-order houses such as Sears, Roebuck. That company and others offered plans and materials

DAIRY FARM ARCHITECTURE

through catalogue sales. The houses listed and illustrated were described as "large but economical, efficient, and able to serve diverse needs of a modern family." While the small and cozy bungalow was becoming popular in the cities and suburbs, this large square, two-story house type became the choice of many farmers, especially in Stearns County, as a dwelling that would give the effect of a manor house on an established country estate.

The Blonigen farm near St. Martin, Minnesota, around 1960. The architectural style and building layout of this farm is characteristic of dairy farms built between the mid-1880s and early 1920s.

During the first decades of the twentieth century, outside influences began to shape the appearance and nature of the dairy farm in Stearns County. Life in rural America became less isolated with the coming of the automobile age and mechanized agriculture. Government involvement in agricultural practices increased through legislated policies. Various professional agencies began to search for ways to improve crops, livestock, and the architecture of the farm. Plans for barns and silos were published by the federal government as well as by commercial lumberyards. Some of these plans became working specifications for the monumental structures that dominate the site of almost every dairy farm in the county. By the 1920s some plan books included designs for model farmhouses as pictured in "Her Majesty, the Dairy Cow." These smaller one-and-a-half-story houses were described as being as up to date as the new designs for barns and out buildings.

50

DAIRY FARM ARCHITECTURE

Split-level and ranch houses began to appear in the countryside after World War II, heralding the fact that agriculture and dairy farming were in step with other large-scale business ventures in the modern industrialized world. An operation as large, complex, and contemporary as the Hansen farm in Rockville Township clearly documents this present stage of development in dairy farming. The large farmhouse on this farmsite is dwarfed by numerous larger outbuildings that house either the big machines needed to work the land or the large dairy herd one needs to maintain such an operation. Old and new barns, machine sheds, multiple grain storage units and silos, and a manure tank extend the space and functions of the farmyard but do not essentially alter the pattern of work followed earlier by the Reisinger family on their farm during the 1850s and 1860s. "Her Majesty" demanded then, as she does now, care, concern, and a continual round of seasonal and daily labors to serve her well so that she in turn may serve the farm family that serves us with the bounty of the dairy farm.

The Hansen family farm near Rockville, Minnesota, in 1961. This farm typifies the new class of high-technology, industrialized farms of the latter part of the twentieth century.

CATALOGUE OF ARTIFACTS

Milk pitcher trophy. *1935. This trophy was awarded to A. G. Bork of Paynesville for the best purebred sire at the Stearns County Fair. H. 9".*

CATALOGUE OF ARTIFACTS

Cowbell. c. 1880. H. 7½", W. 6".

Ox shoes. Oxen are fully matured steers. They were the principle beast of burden of early Stearns County farmers. c. 1870. L. 6".

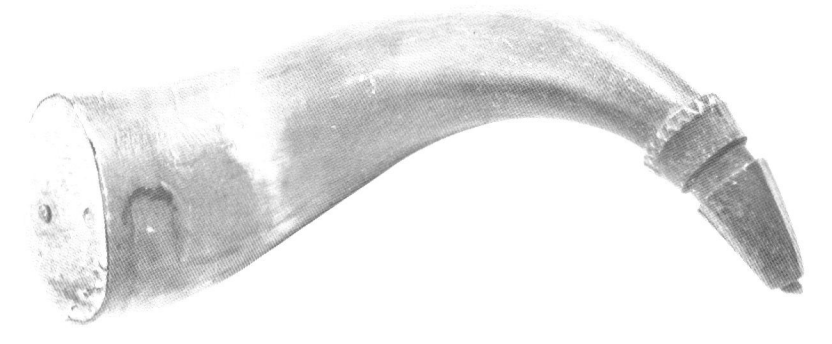

Powder horn. 1862. Cow horns were commonly used as blackpowder containers for nineteenth-century firearms. L. 11¾", Plug diameter 3¼".

Cowbell. c. 1870. Dairy farmers placed cowbells on their livestock so they could find them when it was time for milking. H. 8½", W. 7".

Cast iron horn tip. 1897. Horn tips were screwed onto oxen horns as a safety precaution. H. 2¼", D. 2".

53

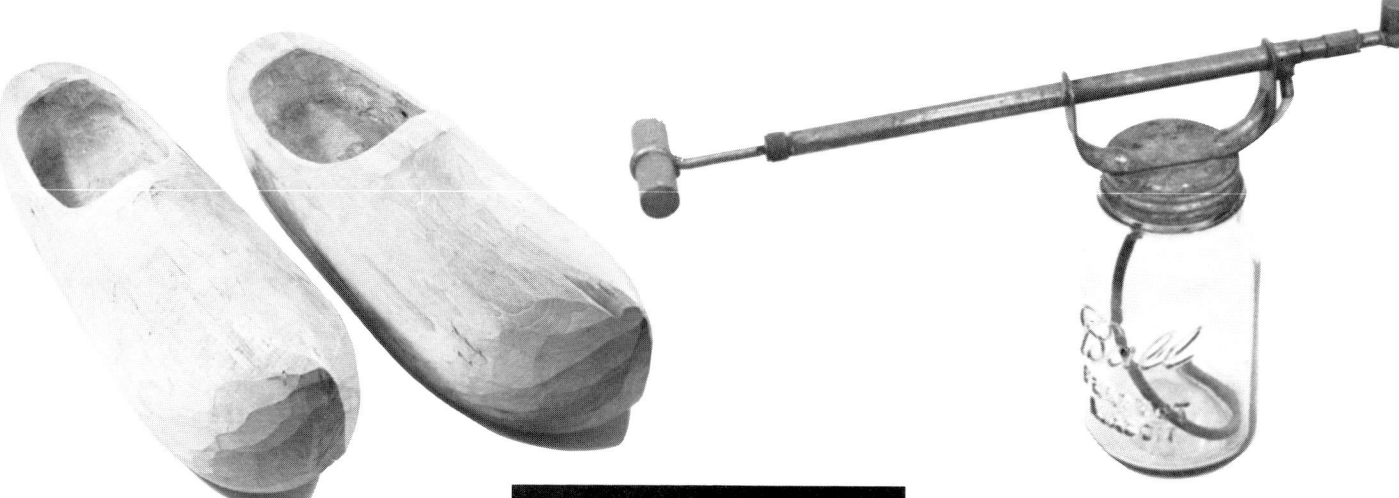

CATALOGUE OF ARTIFACTS

Wooden shoes. *c. 1860.* Many nineteenth-century Stearns County farmers wore wooden shoes inside barns to prevent ruining their leather footwear on the muddy, acidic floors. Wooden shoes also protected their feet when they were stepped on by a cow during milking. L. 1′ 2″, W. 5½″.

Calf weaners. *c. 1900-1920.* Many types of devices were used by farmers to keep calves away from their mothers during weaning time. Left: L. 7″; Right: L. 3½″.

Hudson ceramic salt bowl. *c. 1880.* Salt was given to cows as a mineral supplement. D. 8″, H. 5″.

Cow and livestock insecticide sprayer. *c. 1930.* Insects were a bane to dairymen because they irritated cows and spread disease. H. 8″, L. 10¼″.

Dehorner. *c. 1920-1940.* Farmers removed the horns from their animals to prevent them from injuring other cattle. L. 3′, 1¼″.

54

CATALOGUE OF ARTIFACTS

Commercially manufactured pedestal stool. *c. 1910. This type of stool tipped easily. H. 12", D. 9".*

Homemade wooden milking stool. *c. 1900. H. 11", L. 12¼".*

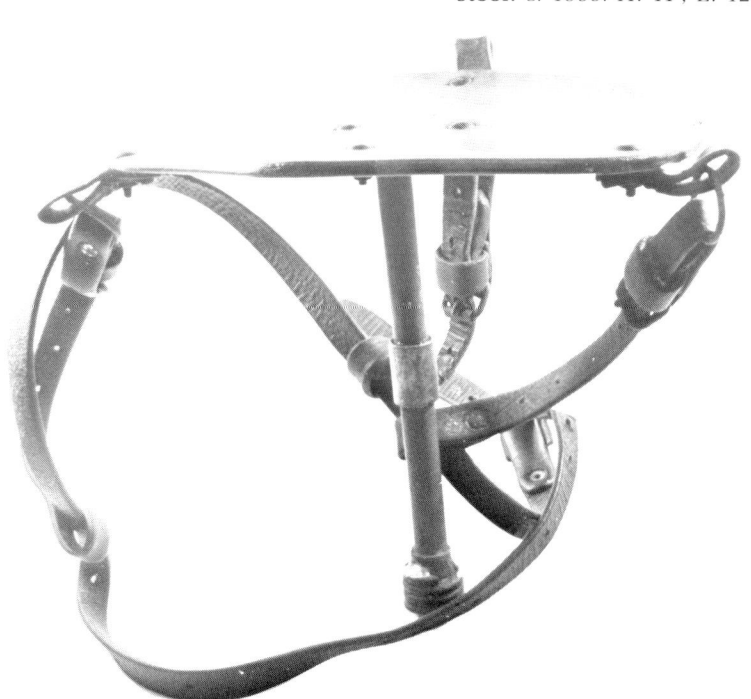

Spring milking stool. *c. 1900. To eliminate the inconvenience of carrying a stool, many farmers purchased ones that could be strapped to the hips. H. 10", W. 12".*

Farm Master Star milking stool. *c. 1935. H. 12", D. 10".*

55

CATALOGUE OF ARTIFACTS

Wooden stave milk bucket with brass bands. *c. 1860.* This bucket was used on the Warnert family farm near St. Joseph, Minnesota. Wooden buckets were difficult to clean and eventually resulted in bad-tasting milk. H. 9″.

Steel milk bucket. *c. 1920-1940.* The seams on this bucket made it difficult to keep clean. H. 10¼″.

Square top milk bucket. *c. 1900.* The shape of the bucket enabled farmers to hold it firmly with their knees while milking the cows. Courtesy of Bert and Gert Schwinghamer. H. 10″.

CATALOGUE OF ARTIFACTS

Hooded milk bucket. *c. 1880.* This bucket was of ideal design for the hand milker. The hood kept debris out of the milk and a bottom handle facilitated easy pouring. H. 13".

Swedish seamless brass milk bucket. *c. 1860.* Brass buckets like this one were highly prized by settlers because they didn't rust and were easy to clean. This pail was brought to St. Cloud by a Swedish immigrant after the Civil War. H. 7", D. 10¾".

Tail holder (top). *c. 1910.* This device was used to secure the cow's tail to its back leg to prevent the milker from being swished in the face. L. 6".

Hobbies (bottom). *c. 1900.* Farmers attached these chains to a cow's rear legs to prevent kicking. L. 2'.

57

CATALOGUE OF ARTIFACTS

Wooden stave churn with dash. *c. 1870. The simple design of this churn is rooted in ancient times. H. 3′ 8″, W. 9½″.*

Number One Belle churn, manufactured in St. Paul, Minnesota. *c. 1890. H. 34″, W. 8″.*

CATALOGUE OF ARTIFACTS

Union Brand diaphragm butter churn, manufactured in St. Louis, Missouri. *1864. H. 33˝, L. 29˝, W. 14˝.*

Metal Dazey churn. *1907. H. 2´, W. 12˝.*

59

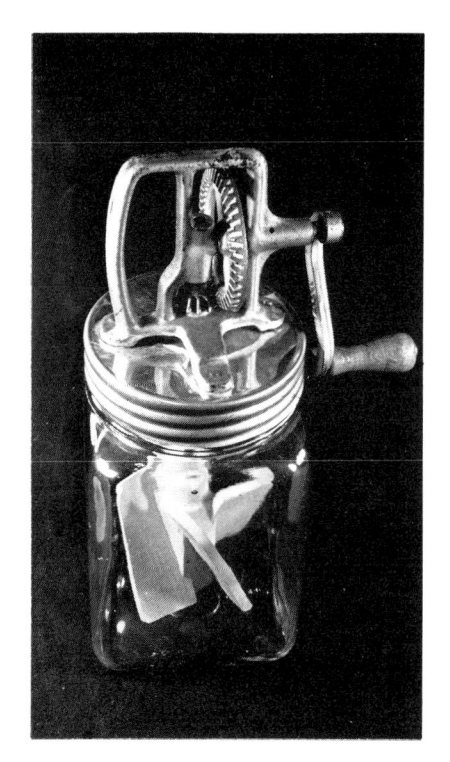

CATALOGUE OF ARTIFACTS

Dazey butter churn. *c. 1900-1950. Small churns like this were common in both rural and urban households. The glass enabled butter makers to see how the butter was progressing. H. 12", W. 4½".*

"New Style" white cedar cylinder churn. *1912. Three-gallon capacity. H. 15½", W. 15".*

Western stoneware butter poultry feeder. *c. 1870. Milk left after butter making (buttermilk) was fed to farm animals. Courtesy of Bert and Gert Schwinghamer. H. 10½", W. 7".*

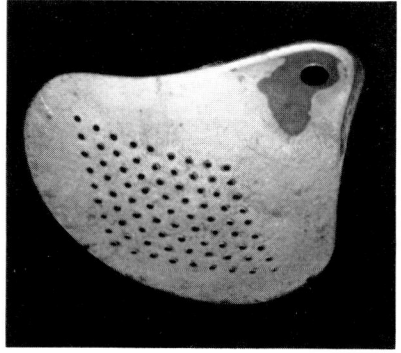

Metal cream skimmer. *c. 1910-1930. This device was used to collect cream from the top of milk for making butter. L. 5¼", W. 5".*

60

CATALOGUE OF ARTIFACTS

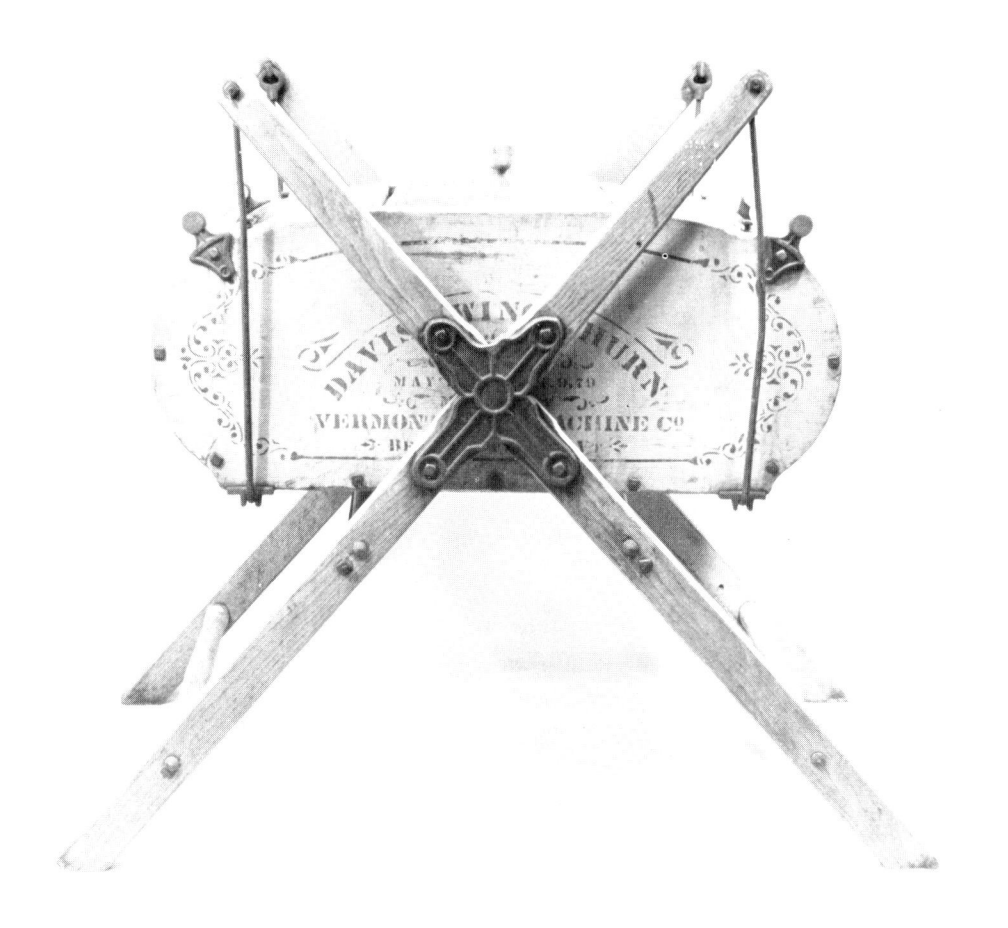

Davis Swing Butter Churn Number Five, manufactured by the Vermont Farm Machine Company, Bellows Falls, Vermont. *This churn was used on the Barthelemy farm near St. Cloud. Children had the task of rocking the churn. H. 3′ 3″.*

61

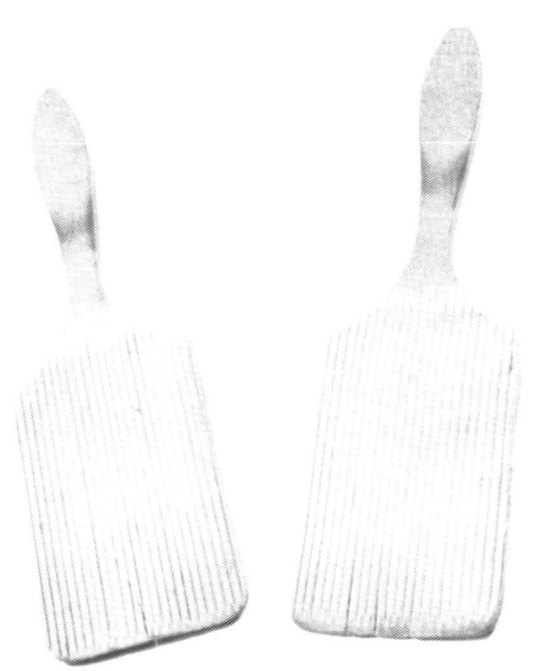

CATALOGUE OF ARTIFACTS

Grooved butter box. *c. 1850. Butter makers used grooved spades to decorate butter. L. 10″, W. 3″.*

Wooden butter bowl with paddle. *c. 1880-1920. After butter was removed from the churn, it was worked in a wooden bowl with paddles such as these to remove excess moisture and work in salt. Bowl H. 3¼″, D. 10½″.*

Hansen's Brand butter color. *1920. Both commercial and home butter makers added artificial color to give butter its distinctive yellow. The natural color of butter varied according to cow breed, season, and type of feed. H. 10-5/8″, W. 5″.*

Butter spades. *c. 1910-1930. Average L. 8″.*

CATALOGUE OF ARTIFACTS

Butter crock. *c. 1920. This one-pound crock was used by the McCrew family on their farm near Watab, Minnesota, in the early part of the twentieth-century. H. 2-3/4", D. 4-5/8".*

Wooden butter molds.
1860-1920. Molds were used by Stearns County butter makers to give distinctive decorative patterns to their butter. They were often homemade. Average. W. 5", H. 4".

German butter box. *c. 1850. This wooden container was brought to the United States by the Breer family, settlers in Richmond, Minnesota. H. 2-5/8", L. 5-1/2", W. 3-3/4".*

63

CATALOGUE OF ARTIFACTS

Five-gallon milk or cream can. *c. 1870-1900. Farmers transported their cream to creameries in cans such as this one. The riveted seams and handles of the early milk cans allowed dirt to accumulate, eventually giving milk a bad flavor. H. 1′ 8″, W. 10″.*

Superior Sealed twelve-quart cream can. *c. 1890. The inverted can cover and narrow neck forced air out of this style of can to prevent churning during transit. H. 12½″, W. 8¾″.*

Red Wing Union Stoneware five-gallon crock. *c. 1900-1910. This crock was used as a cream can on the Dullinger farm near St. Joseph. H. 1′ 5″, W. 11″.*

CATALOGUE OF ARTIFACTS

Seamless creamery milk cans. c. 1930–1970. By the middle of the twentieth century, Stearns County dairy farmers were shipping whole milk in standard eight- to ten-gallon cans to creameries and centralizers. The bands around the shoulder and bottom of the cans prevented them from being battered in transit. Average H. 2′, D. 1′1″.

Milk can filter holder. c. 1910. This wire holder was given away as a premium at Loso's Hardware Store in St. Joseph, Minnesota. D. 2′4″.

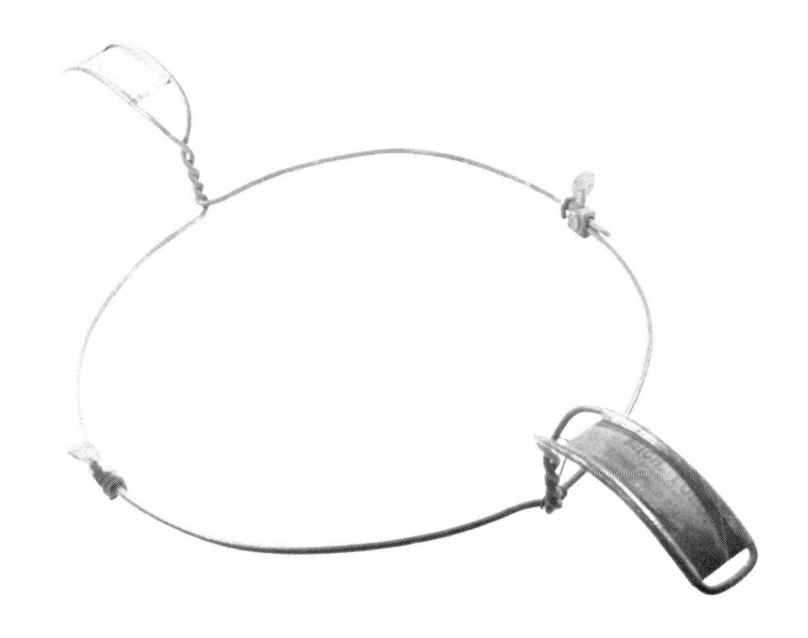

65

CATALOGUE OF ARTIFACTS

Milk strainer cover. c. 1930-1940. In the 1930s, the federal government required covers on all milk-receiving equipment to prevent contamination by insects. This cover was made by a blacksmith in Albany, Minnesota. D. 1′ 1¾″.

DeLaval Factory cream separator. 1930-1960. Separators reputed to "skim milk so clean that but a small fraction of one percent of butterfat would remain in the milk" were first introduced in Stearns County in the 1890s. Factory cream separators allowed farmers to bring in their milk to creameries to be mechanically separated. Once it was separated, the creameries bought the cream and the farmers took skim milk home for their animals. This separator was used at the Meire Grove Creamery. H. 4′ 10″, L. 8′ 8″.

Shotgun cream can. c. 1900-1930. This style of can was originally meant to transport cream. In later years families used them to cool milk. H. 1′ 2½″, W. 9″.

Milk strainers. Strainers were used to remove large, foreign particles from raw milk. Early strainers (top, c. 1890) had brass screens to avoid rust. H. 7″, D. 11″. Later strainers (bottom, c. 1950) utilized removable paper filters held in place by a wire. H. 8″, D. 12½″.

CATALOGUE OF ARTIFACTS

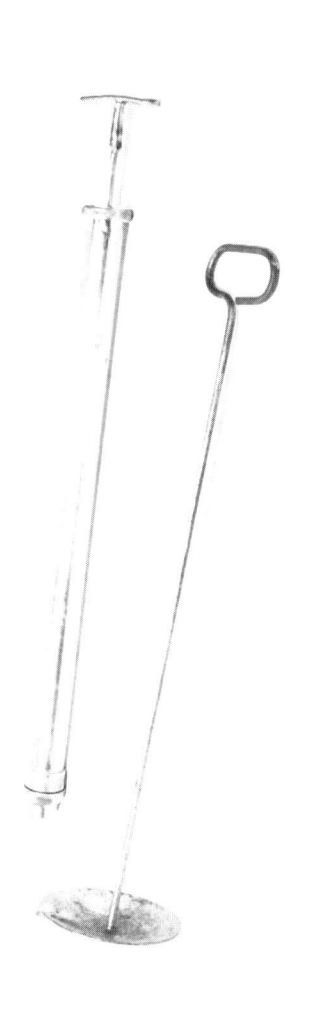

Official Babcock Test centrifuge. 1912-1924. Many dairy farmers purchased small testers to check their milk. This two-tube model was manufactured by the Creamery Package Company, Chicago. H. 7".

Babcock testing apparatus. 1950-1970. Creameries paid farmers according to the amount of butterfat in the milk they brought in. Early methods of measurement based on the natural separation of cream were crude and led to disputes between farmers and butter makers. In 1890, Stephen A. Babcock solved the problem by introducing a reliable, scientific means of measurement called the "Babcock Test." The process was fast, simple, and accurate: sulfuric acid was added to a milk sample to break down the solid fat, then placed in a special test bottle and whirled in a centrifuge. The fat gathered in the narrow graduated neck of the test bottle where it could be seen and measured. Top to bottom: Cream testing bottle. H. 6", Skim milk testing bottle. H. 5", Calipers. L. 5", Acid dipper. L. 1'4", Glass acid dipper. 8", Pipette. L. 11½".

Milk sediment tester. *(Left)* c. 1950-1960. A creamery's reputation rested on how well its milk products tested for contaminants. The main source of bad-tasting milk was foreign sediments. Butter makers at the Meire Grove Creamery used this device to extract milk samples from the bottom of milk cans for testing. L. 2'3".

Milk testing ladle. *(Right)* 1960. To obtain representative milk samples to test for butterfat content, butter makers used ladles such as this one. It was used at the Albany Creamery. L. 2'1".

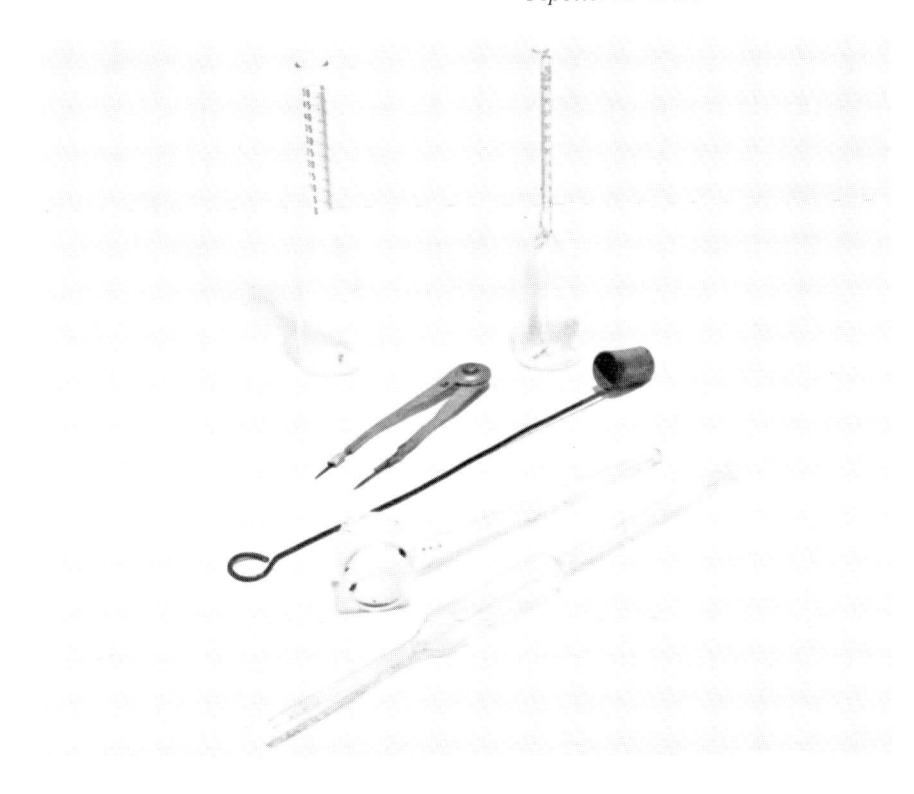

67

CATALOGUE OF ARTIFACTS

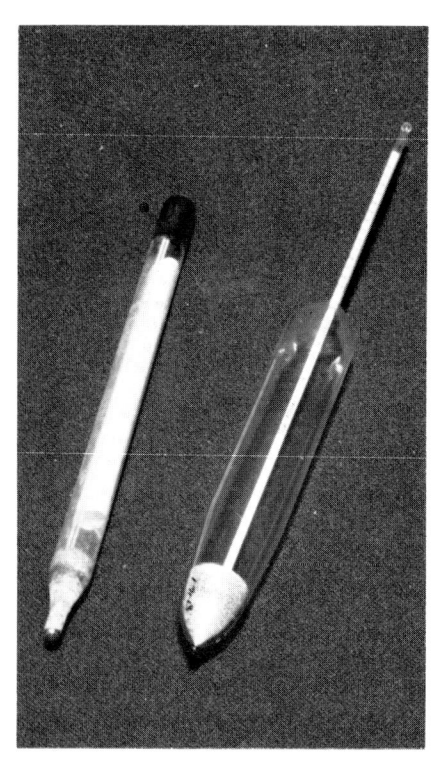

Pasteurizing thermometers.
Milk was routinely heated by creameries to kill potentially harmful organisms. Left: L. 8", Right: L. 12".

Babcock Test centrifuge. *c. 1930.*

CATALOGUE OF ARTIFACTS

Improved Jalco Babcock Test centrifuge. *c. 1940-1960. This electric tester was used at the St. Martin Coop Creamery. It had a thirty-six bottle capacity. D. 1′ 11″, H. 1′ 6½″.*

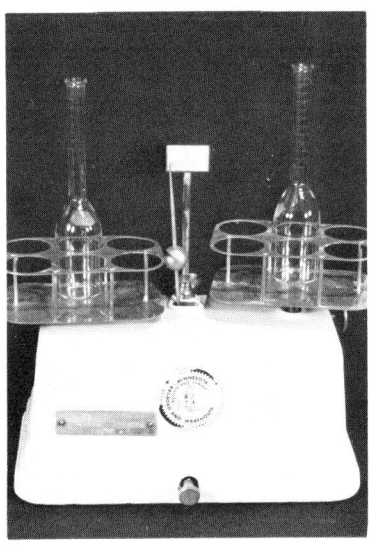

Torsion cream test scale. *1950-1970. Scales were used along with the Babcock Test to determine amount of moisture in milk. This scale was used at the Meire Grove Creamery. L. 10″, H. 10″.*

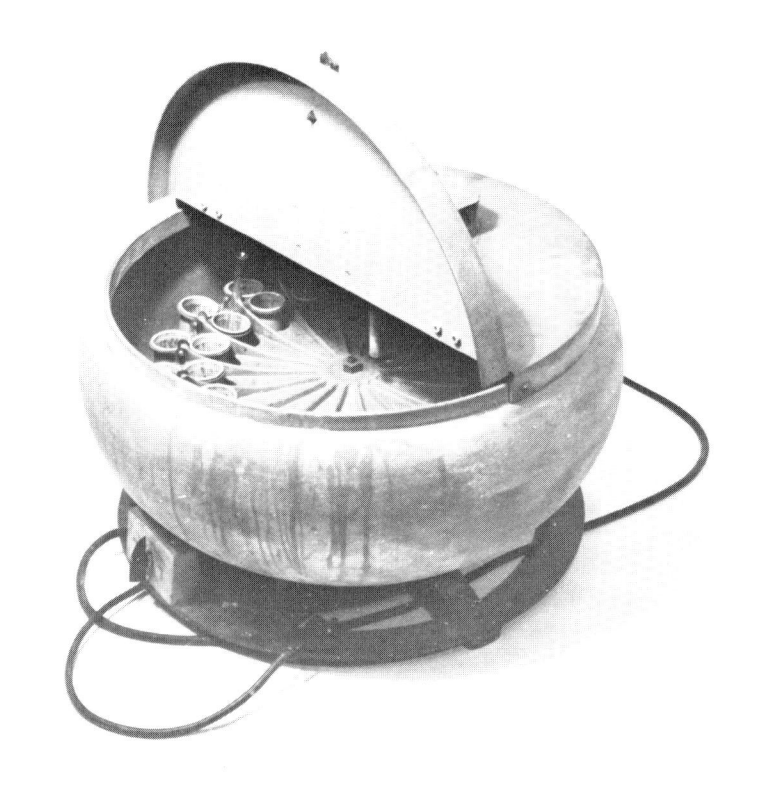

Creamery Package Manufacturing Company eight-bottle twentieth-century Babcock Test centrifuge. *c. 1912.*

69

CATALOGUE OF ARTIFACTS

Creamery butter ladles and packer. *c. 1940-1960. These instruments were used by butter makers at the Cold Spring Creamery to handle and pack bulk butter. Top to Bottom. Packer. H. 1'5", W. 5". Ladle. 1'3", W. 8". Ladle. L. 10", W. 6".*

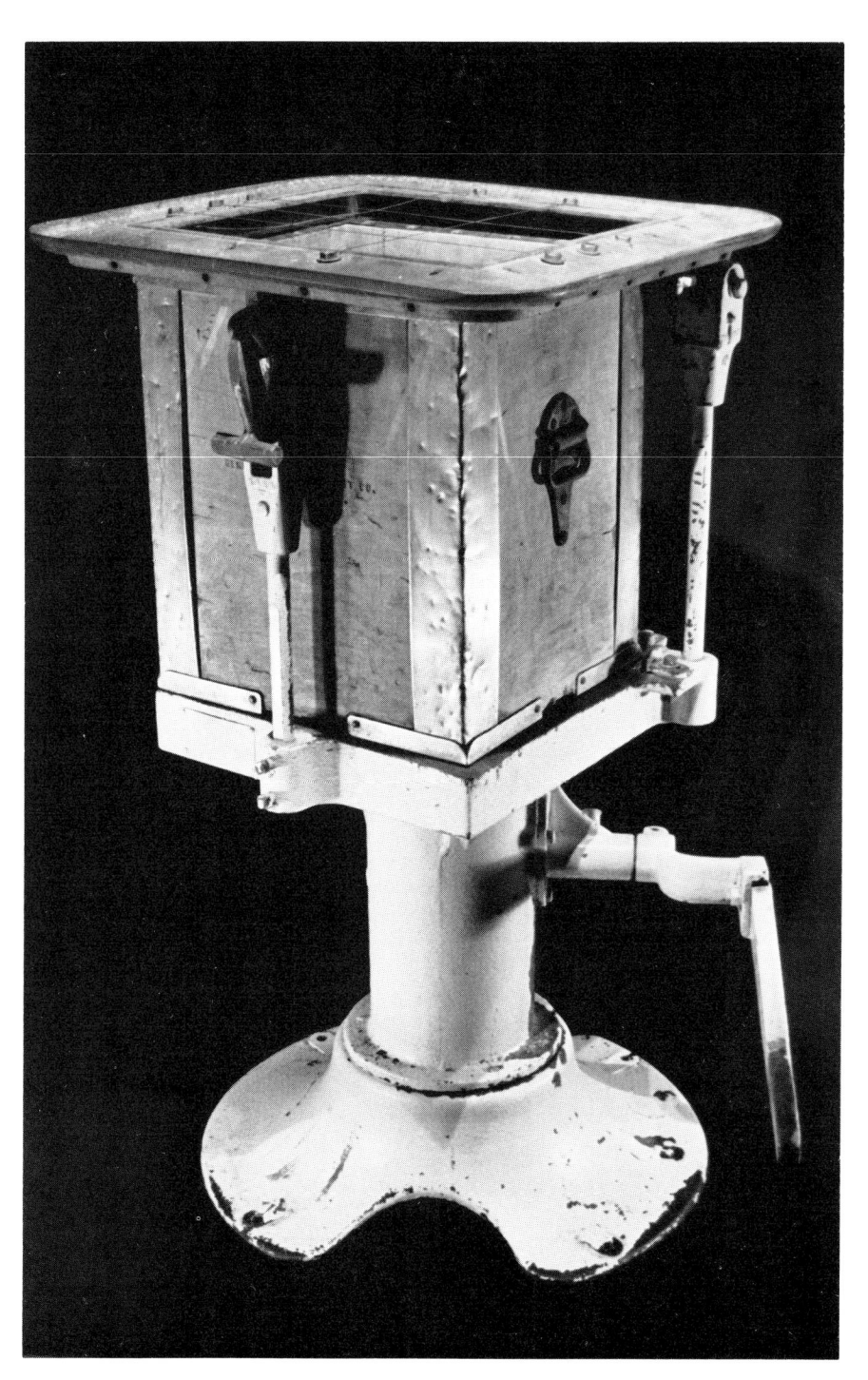

Lincoln butter cutter. *c. 1920. The butter cutter or "Printer" enabled butter makers at coop creameries to mass produce one-pound blocks of butter. The process was simple: Fresh butter was packed into wooden "Friday" boxes with stompers and ladles to eliminate air pockets. As the false bottom in the box was raised with a lever, the butter was cut into rectangles. A hand-held wire cutter was drawn through the butter horizontally to separate the blocks. The butter was then ready for packaging and final weighing. When the box emptied, it was replaced with a full one. This cutter was used at the Cold Spring Creamery. H. 3'6".*

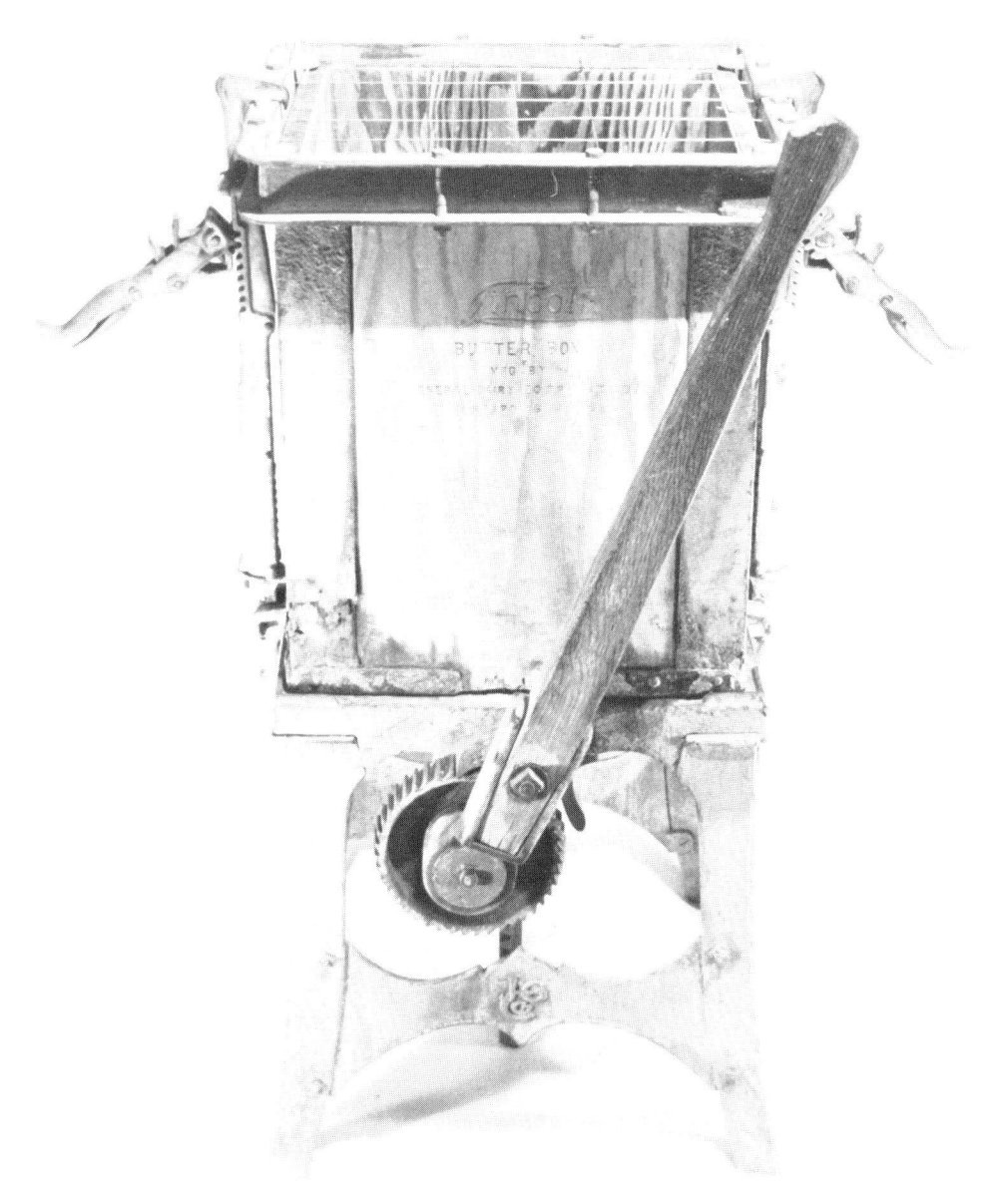

CATALOGUE OF ARTIFACTS

J. C. Cherry Friday butter printer. *c. 1930. This butter cutter was used at the Cold Spring Creamery. H. 3′ 1½″.*

71

CATALOGUE OF ARTIFACTS

Mojonnier butter print scale. *1917. Scales were used to weigh individual pounds of butter to check for accuracy. This scale was rejected by the Minnesota Division of Weights and Measures. Meire Grove Creamery. H. 1′ 4″, L. 1′ 3″.*

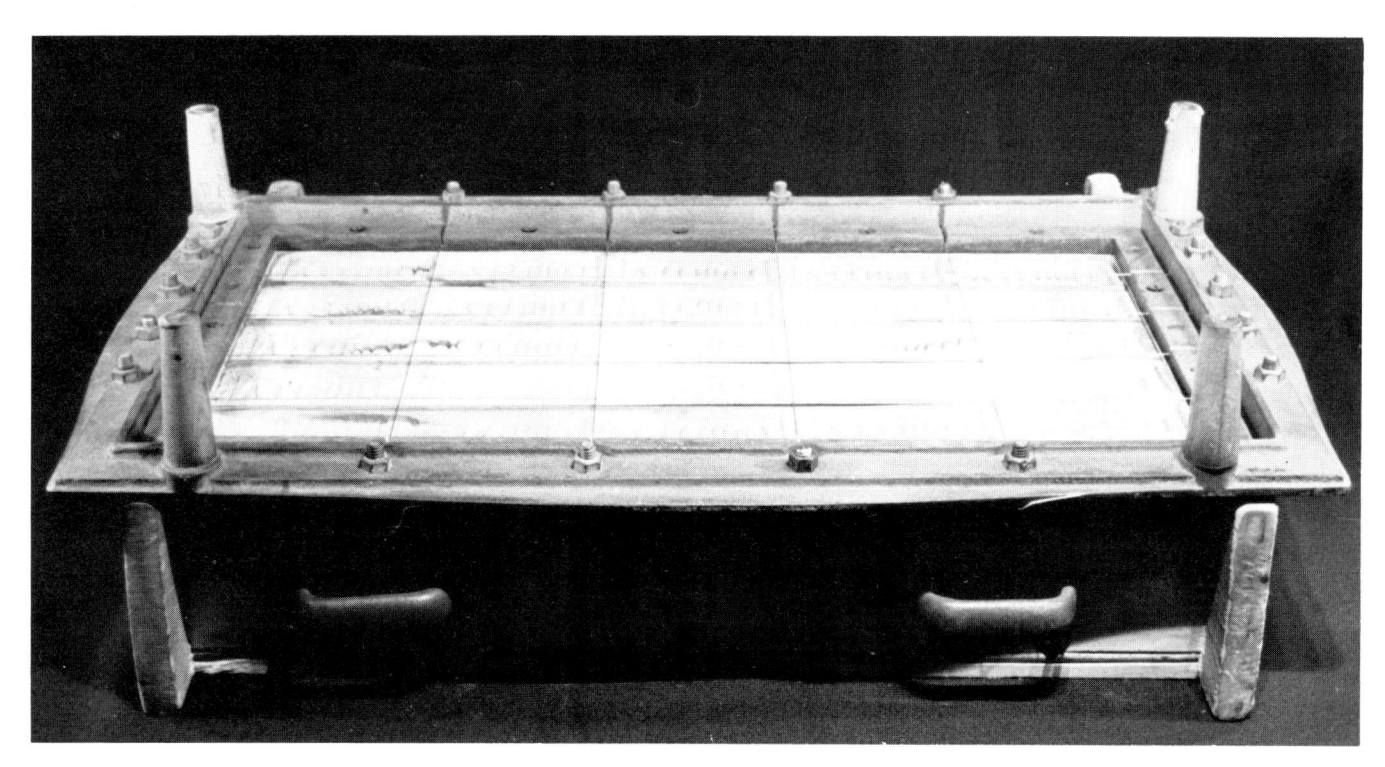

Butter cutter. *c. 1940. "Louella" Brand Butter was made at the Meire Grove Creamery with this cutter. L. 2′ 5″, H. 6″, W. 1′ 9″.*

CATALOGUE OF ARTIFACTS

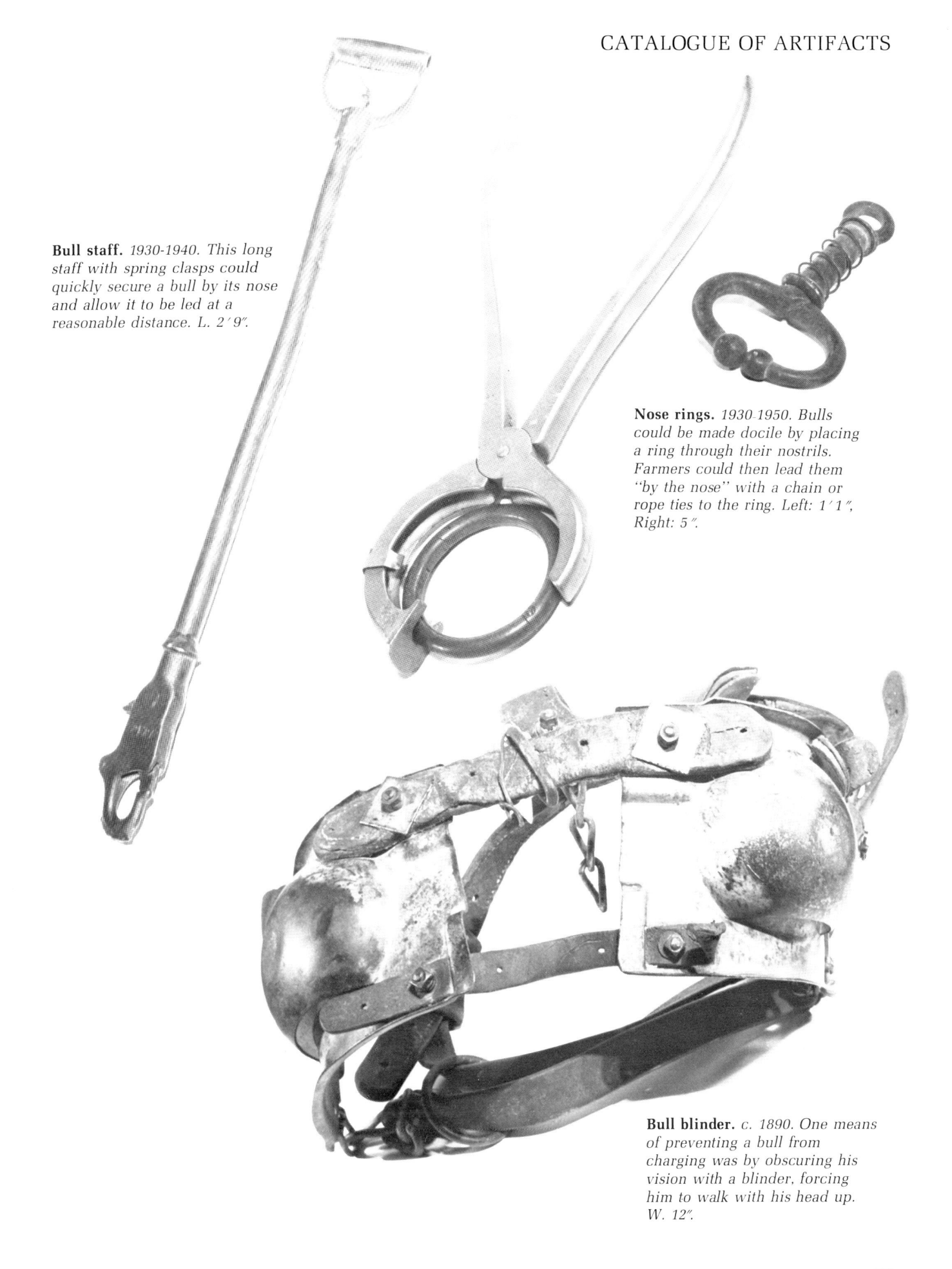

Bull staff. *1930-1940. This long staff with spring clasps could quickly secure a bull by its nose and allow it to be led at a reasonable distance. L. 2′9″.*

Nose rings. *1930-1950. Bulls could be made docile by placing a ring through their nostrils. Farmers could then lead them "by the nose" with a chain or rope ties to the ring. Left: 1′1″, Right: 5″.*

Bull blinder. *c. 1890. One means of preventing a bull from charging was by obscuring his vision with a blinder, forcing him to walk with his head up. W. 12″.*

73

CATALOGUE OF ARTIFACTS

Artificial insemination equipment. *1937. Artificial insemination of cows became commonplace by the late 1950s and eliminated the need for herd bulls. This equipment was used by Bert Schwinghamer, a dairy farmer near Albany, Minnesota. Left to right: Artificial vagina made from an inner-tube from a Model T auto tire, used to collect semen, L. 1′ 8″. Speculum tube, L. 1′ 3½″. Gelatin capsules for holding bull semen, L. 1½″. Pipette used to insert semen through speculum tube into cow, L. 1′ 5½″.*

Veterinarian's equipment. *1920-1940. As more farmers "turned to dairying," they came to rely on veterinarians to ensure the health of their herds. Clockwise left to right: Medicine baller, L. 1′ 4″. Emasculator, L. 1′ 2″. Syringe kit, L. 11″, W. 3½″. Mortar and pestle, H. 3″, D. 5″. Udder balm, D. 4½″.*

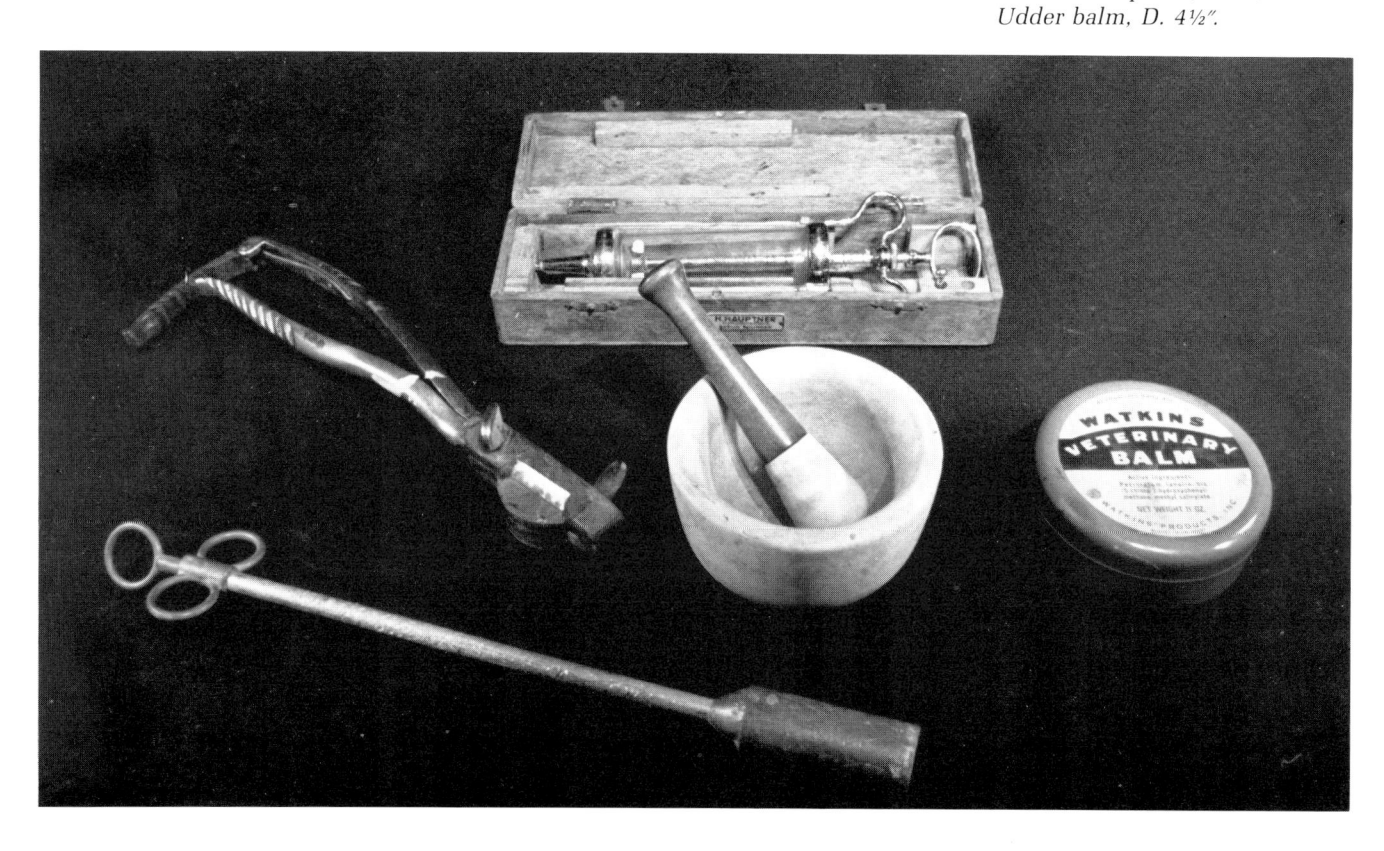

CATALOGUE OF ARTIFACTS

Settling can. *c. 1880. Settling cans were the forerunner of mechanical cream separators and served farmers as an early means of measuring cream. Milk was poured into the can and allowed to cool so the cream (butterfat particles are lighter than milk) would rise. Windows on the side of the can indicated the amount of cream. Once the separation was complete, milk was drained from the bottom through a spigot. H. 16″, W. 13″.*

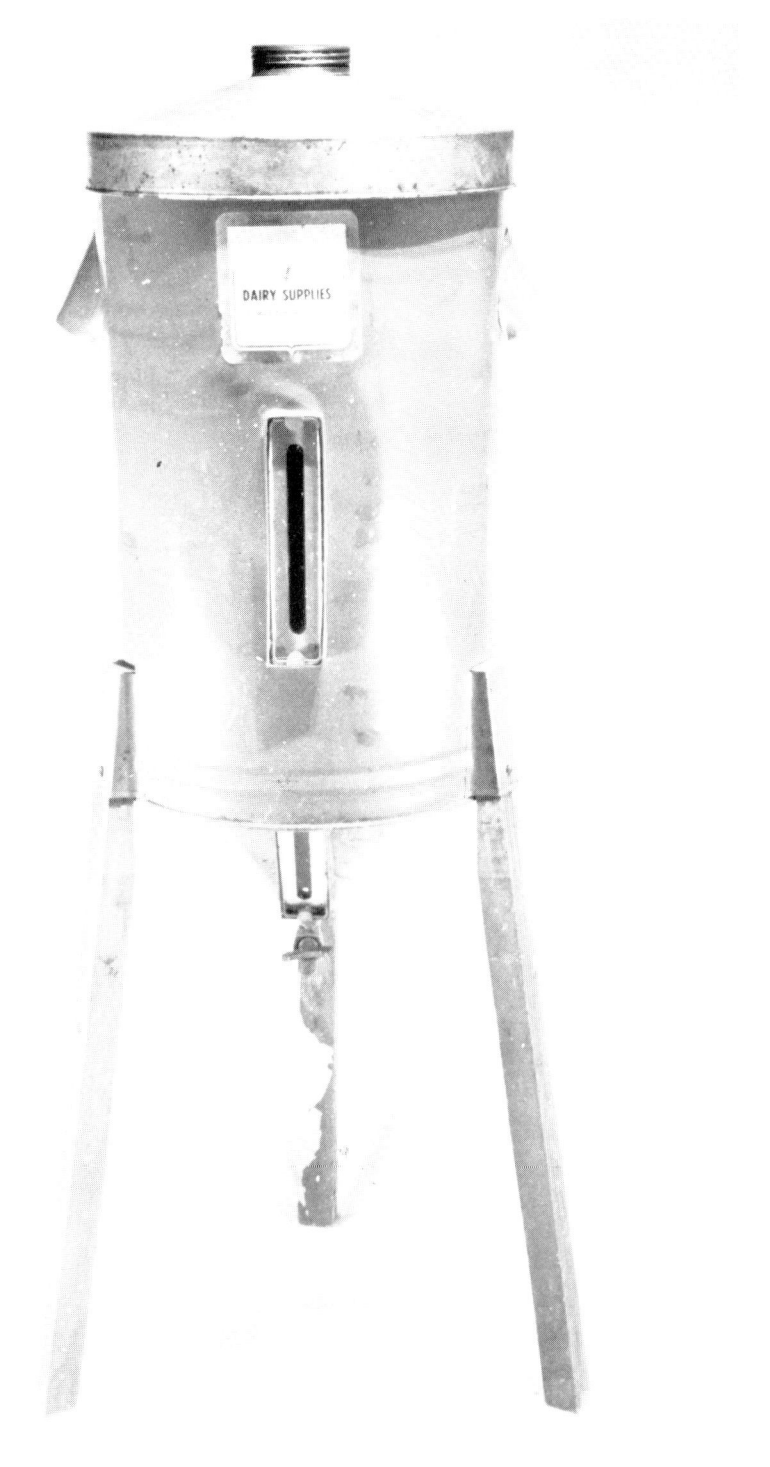

Superior ten-gallon settling can. *c. 1910. H. 3′4″, D. 1′1″.*

CATALOGUE OF ARTIFACTS

Anker Hoth self-balancing #6 cream separator. *1917.* In 1878, the first mechanical cream separators were introduced in America by France's DeLaval Company. The new device offered a quick and efficient means of separating cream and eliminated the need for farmers to haul whole milk to creameries. By the beginning of World War I, the machines were a common fixture on dairy farms, despite the painstaking task of cleaning them after each use. H. 3′ 9″.

CATALOGUE OF ARTIFACTS

Melotte cream separator. *c. 1920. This manually operated separator was manufactured in Belgium. H. 3′ 11½″.*

CATALOGUE OF ARTIFACTS

DeLaval cream separator #519.
c. 1935. By the 1920s, optical electric motors appeared on new models of separators. H. 3′9″.

CATALOGUE OF ARTIFACTS

DeLaval Junior cream separator. *c. 1920. Small, lightweight, tabletop separators were marketed by DeLaval and other companies for personal use in town and country homes. H. 1′ 7″.*

CATALOGUE OF ARTIFACTS

Cream separator oils. *c. 1950-1960. Cream seperators had to be kept immaculately clean and oiled. DeLaval, H. 7½". Superla, H. 10".*

McCormick power washing cream separator. *1960. This separator boasted a revolutionary feature—an automatic disk washer, eliminating the chore of cleaning the machine by hand after each use. H. 3′ 7".*

CATALOGUE OF ARTIFACTS

McCormick-Deering single cow milking machine. *c. 1950. H. 1′ 8″.*

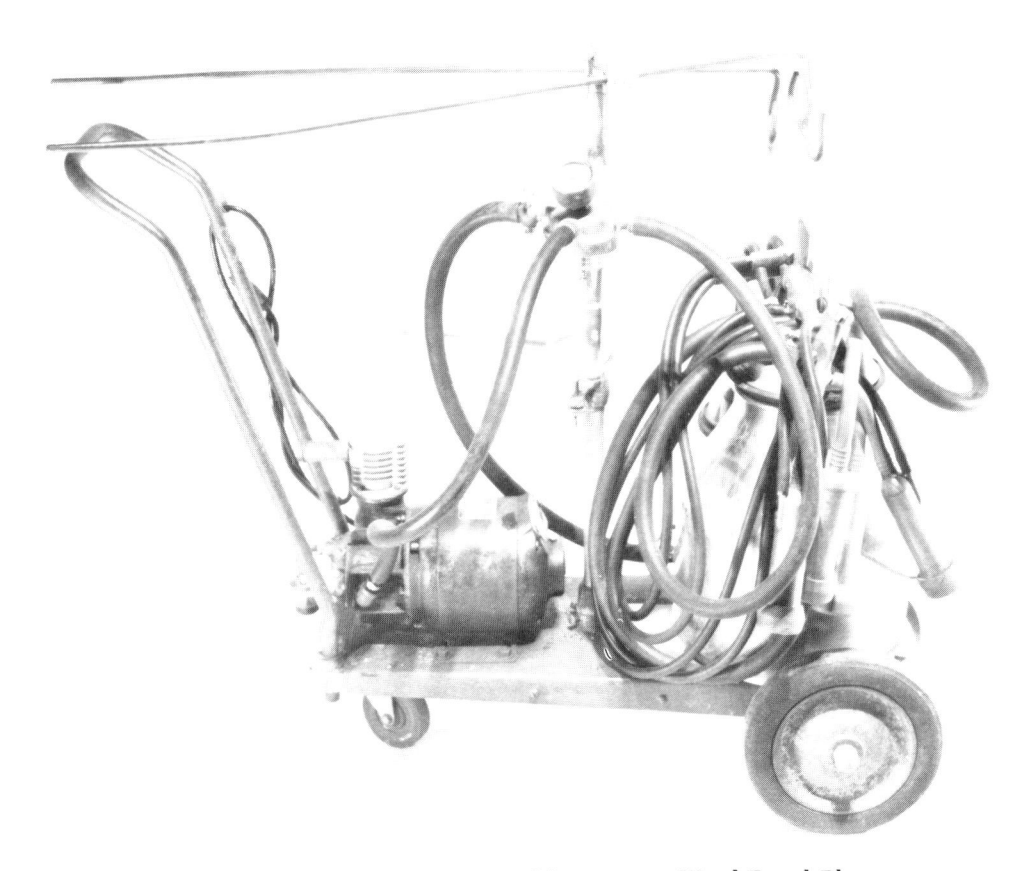

Montgomery Ward Royal Blue two-cow milking machine. *1932. This electric model boasted a three-wheeled cart that could be easily moved down the barn aisle. L. 3′ 3″, H. 3′ 1″.*

Surge milker. *c. 1960. Surge milkers were hung below the cow with a leather belt. Their stainless steel make and lack of seams made them easy to clean and very popular in Stearns County. D. 15¾″, H. 8″.*

81

CATALOGUE OF ARTIFACTS

Organic two-cow milking machine. *1942.* Milking machines came into common use in Stearns County in the 1930s and 1940s with the introduction of rural electricity. Prior to electrical power, milking machines were run with gasoline engines. The new machines cut milking time by half and enabled dairy farmers to double herd sizes. This machine was used until 1960 by Claude Dullinger, near St. Joseph, Minnesota; he affectionately referred to it as "George." H. 2″, W. 1′.

Universal Coop milker. *c. 1935.* This electric-powered compressor created the vacuum needed to power milking machines. L. 1′10″, H. 1′9″.

DeLaval milking machine bucket. *c. 1910.* This early milker bucket was made completely of brass. Courtesy Joe Miller. H. 1′6″.

CATALOGUE OF ARTIFACTS

DHIA-approved milkometer. *c. 1965. Milk meters gave dairy farmers a means to accurately measure the amount of milk each cow produced each day. L. 1′1″, W. 9″.*

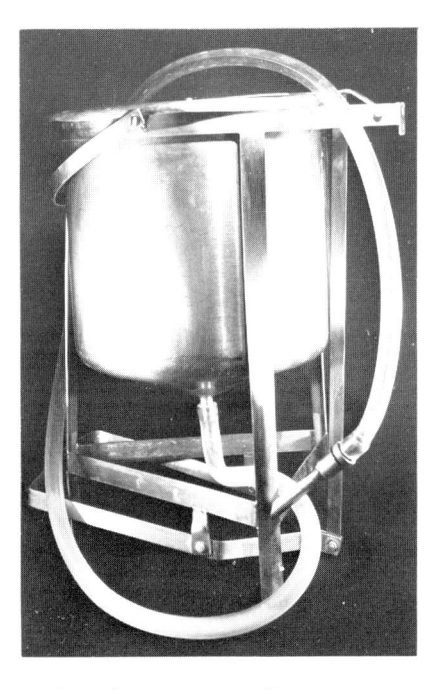

Step-Savers. *1965. Step-savers were connected directly from bulk tanks with hoses to the cows in the barn to receive milk. They eliminated the need for dairy farmers to walk all the way to the milk room to dump milk into the bulk tank after each milking. H. 2″, D. 1′3″.*

CATALOGUE OF ARTIFACTS

Wooden stave cooling tank. c. 1920-1960. Prior to the advent of electrical power, milk cans were cooled with running water. This tank was filled with water pumped in by a windmill. H. 2′6″, D. 4′1″.

Sunset milk cooler. c. 1960. Stainless steel bulk tanks eliminated the use of milk cans and enabled automatic control of milk temperature. This tank was manufactured in St. Paul. L. 6′, W. 3′5″, H. 4′.

CATALOGUE OF ARTIFACTS

Four-quart cream pail. c. 1880. Towns people in the late nineteenth and early twentieth centuries commonly bought their milk from nearby farmers and carried it home in containers such as this one. H. 9½".

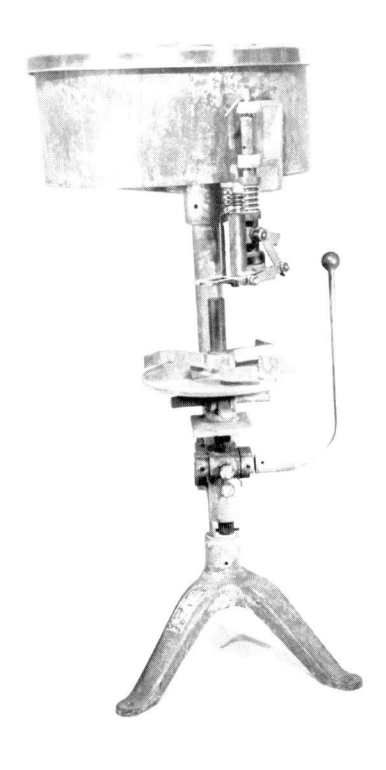

White enamel two-gallon cream pail. c. 1925. Small one- or two-gallon pails were used to collect cream or milk from separators for family use. H. 11", W. 8¼".

Fort Wayne Dairy Equipment Company manually operated bottle machine. 1915-1930. Some families operated small dairies that bottled and sold milk products to townspeople. This bottler was used by the Borgmann Family Dairy near Sauk Centre. The small dairy operated from 1914 to 1976. H. 4′.

85

Cream top milk bottle. *c. 1920. This distinctively designed bottle allowed cream to collect in the top bulb. A cream spoon was used to pinch off the milk to pour off the cream. H. 9¼".*

Insulated route box. *c. 1940-1990. Boxes such as this one were once found on almost every porch in town that received milk products delivered by milkmen. H. 11¼", W. 12", L. 15".*

Milk bottle carrier. *c. 1915. L. 17½", H. 14".*

CATALOGUE OF ARTIFACTS

Milk and cream bottles.
1930-1960. Glass milk bottles were patented in the 1880s and dominated milk distribution until the 1950s. Tallest bottle, H. 9", Smallest bottle, H. 3¼".

Pewter milk bottle cap opener. *c. 1950.* **Wooden hand stomper.** *c. 1940.* Hand stompers were used to apply caps to glass bottles. Stomper, D. 1¼". Cap opener, L. 5".

Stearns County Coop Creamery one-pound butter containers. *1955-1970.* The Cold Spring Creamery produced the last coop creamery butter in Stearns County in 1986. L. 4", H. 2½".

Cold Spring Creamery milk carton. *1970.* Wax or plastic-coated paper "cartons" made up 70 percent of all commercial milk containers by 1967. H. 10".

CATALOGUE OF ARTIFACTS

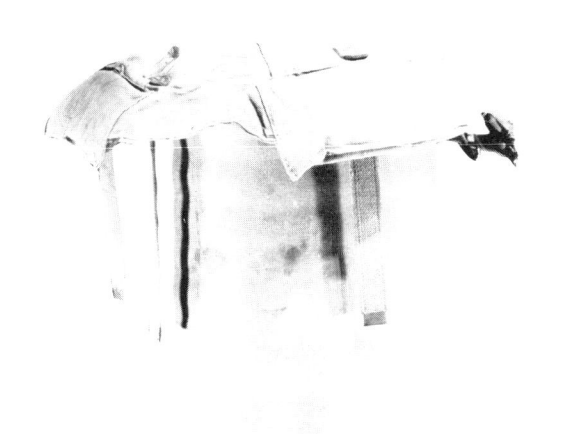

Insulated ice cream shipper. *c. 1950. Creameries that made ice cream, like the St. Martin Coop, delivered their products to grocery stores in containers such as this one. H. 1′ 11″, D. 11″.*

Ice cream freezers. *c. 1900-1930. Making ice cream at home was a special treat for farm families. Right: Metal two-quart freezer, H. 9″. Left: Fre-zee-zee freezer, four-quart, H. 1′ 3″.*

Insulated ice cream pail. *c. 1930-1940. This two-gallon pail was used by Quality Dairy in St. Cloud, Minnesota. H. 11″.*

Metal ice cream container. *c. 1900. This container was used to deliver brick ice cream in the Jones Candy Store in St. Cloud. L. 6″, H. 2½″.*